T0212291

Fault Tolerant Computer Architecture

Synthesis Lectures on Computer Architecture

Editor
Mark D. Hill, University of Wisconsin, Madison

Synthesis Lectures on Computer Architecture publishes 50 to 150 page publications on topics pertaining to the science and art of designing, analyzing, selecting and interconnecting hardware components to create computers that meet functional, performance and cost goals.

© Springer Nature Switzerland AG 2022
Reprint of original edition © Morgan & Claypool 2009

All rights reserved. No part of this publication may be reproduced, stored in a retrieval system, or transmitted in any form or by any means—electronic, mechanical, photocopy, recording, or any other except for brief quotations in printed reviews, without the prior permission of the publisher.

Fault Tolerant Computer Architecture
Daniel Sorin

ISBN: 978-3-031-00595-4 paperback

ISBN: 978-3-031-01723-0 ebook

DOI: 10.1007/978-3-031-01723-0

A Publication in the Springer series

SYNTHESIS LECTURES ON COMPUTER ARCHITECTURE

Lecture #5

Series Editor: Mark D. Hill, University of Wisconsin, Madison

Series ISSN
ISSN 1935-3235   print
ISSN 1935-3243   electronic

Fault Tolerant Computer Architecture

Daniel J. Sorin
Duke University

SYNTHESIS LECTURES ON COMPUTER ARCHITECTURE #5

ABSTRACT

For many years, most computer architects have pursued one primary goal: performance. Architects have translated the ever-increasing abundance of ever-faster transistors provided by Moore's law into remarkable increases in performance. Recently, however, the bounty provided by Moore's law has been accompanied by several challenges that have arisen as devices have become smaller, including a decrease in dependability due to physical faults. In this book, we focus on the dependability challenge and the fault tolerance solutions that architects are developing to overcome it. The two main purposes of this book are to explore the key ideas in fault-tolerant computer architecture and to present the current state-of-the-art—over approximately the past 10 years—in academia and industry.

KEYWORDS

fault tolerance (or fault tolerant), reliability, dependability, computer architecture, error detection, error recovery, fault diagnosis, self-repair, autonomous, dynamic verification

Dedication

"To Deborah, Jason, and Julie"

Acknowledgments

I would like to thank my family for their support while I was writing this lecture. I would also like to thank Mark Hill for inviting me to write this lecture and Mike Morgan for organizing the production of the lecture. Valuable feedback on early drafts of the lecture was provided by Babak Falsafi, Jude Rivers, and Mark Hill. I would also like to thank Lihao Xu for helping me with a question about error coding.

Contents

CHAPTER 1

Introduction

For many years, most computer architects have pursued one primary goal: performance. Architects have translated the ever-increasing abundance of ever-faster transistors provided by Moore's law into remarkable increases in performance. Recently, however, the bounty provided by Moore's law has been accompanied by several challenges that have arisen as devices have become smaller, including a decrease in dependability due to physical faults. In this book, we focus on the dependability challenge and the fault tolerance solutions that architects are developing to overcome it.

The goal of a fault-tolerant computer is to provide *safety* and *liveness*, despite the possibility of faults. A *safe* computer never produces an incorrect user-visible result. If a fault occurs, the computer hides its effects from the user. Safety alone is not sufficient, however, because it does not guarantee that the computer does anything useful. A computer that is disconnected from its power source is safe—it cannot produce an incorrect user-visible result—yet it serves no purpose. A *live* computer continues to make forward progress, even in the presence of faults. Ideally, architects design computers that are both safe and live, even in the presence of faults. However, even if a computer cannot provide liveness in all fault scenarios, maintaining safety in those situations is still extremely valuable. It is preferable for a computer to stop doing anything rather than to produce incorrect results. An often used example of the benefits of safety, even if liveness cannot be ensured, is an automatic teller machine (ATM). In the case of a fault, the bank would rather the ATM shut itself down instead of dispensing incorrect amounts of cash.

1.1   GOALS OF THIS BOOK

The two main purposes of this book are to explore the key ideas in fault-tolerant computer architecture and to present the current state-of-the-art—over approximately the past 10 years—in academia and industry. We must be aware, though, that fault-tolerant computer architecture is not a new field. For specific computing applications that require extreme reliability—including medical equipment, avionics, and car electronics—fault tolerance is always required, regardless of the likelihood of faults. In these domains, there are canonical, well-studied fault tolerance solutions, such as triple modular redundancy (TMR) or the more general N-modular redundancy (NMR) first proposed by von Neumann [45]. However, for most computing applications, the price of such heavyweight, macro-scale redundancy—in terms of hardware, power, or performance—outweighs its benefits, particularly when physical faults are relatively uncommon. Although this book does not delve into the details of older systems, we do highlight which key ideas originated in earlier systems. We strongly encourage interested readers to learn more about these historical systems, from both classic textbooks [27, 36] and survey papers [33].

1.2   FAULTS, ERRORS, AND FAILURES

Before we explore how to tolerate faults, we must first understand the faults themselves. In this section, we discuss faults and their causes. In Section 1.3, we will discuss the trends that are leading to increasing fault rates.

We consider a *fault* to be a physical flaw, such as a broken wire or a transistor with a gate oxide that has broken down. A fault can manifest itself as an *error*, such as a bit that is a zero instead of a one, or the effect of the fault can be *masked* and not manifest itself as any error. Similarly, an error can be masked or it can result in a user-visible incorrect behavior called a *failure*. Failures include incorrect computations and system hangs.

1.2.1   Masking

Masking occurs at several levels—such as faults that do not become errors and errors that do not become failures—and it occurs because of several reasons, including the following.

Logical masking. The effect of an error may be logically masked. For example, if a two-input AND gate has an error on one input and a zero on its other input, the error cannot propagate and cause a failure.

Architectural masking. The effect of an error may never propagate to architectural state and thus never become a user-visible failure. For example, an error in the destination register specifier of a NOP instruction will have no architectural impact. We discuss in Section 1.5 the concept of *architectural vulnerability factor* (AVF) [23], which is a metric for quantifying what fraction of errors in a given component are architecturally masked.

Application masking. Even if an error does impact architectural state and thus becomes a user-visible failure, the failure might never be observed by the application software running on the processor. For example, an error that changes the value at a location in memory is user-visible; however, if the application never accesses that location or writes over the erroneous value before reading it again, then the failure is masked.

Masking is an important issue for architects who are designing fault-tolerant systems. Most importantly, an architect can devote more resources (hardware and the power it consumes) and effort (design time) toward tolerating faults that are less likely to be masked. For example, there is no need to devote resources to tolerating faults that affect a branch prediction. The worst-case result of such a fault is a branch misprediction, and the misprediction's effects will be masked by the existing logic that recovers from mispredictions that are *not* due to faults.

1.2.2 Duration of Faults and Errors

Faults and errors can be *transient*, *permanent*, or *intermittent* in nature.

Transient. A transient fault occurs once and then does not persist. An error due to a transient fault is often referred to as a *soft error* or *single event upset*.

Permanent. A permanent fault, which is often called a *hard fault*, occurs at some point in time, perhaps even introduced during chip fabrication, and persists from that time onward. A single permanent fault is likely to manifest itself as a repeated error, unless the faulty component is repaired, because the faulty component will continue to be used and produce erroneous results.

Intermittent. An intermittent fault occurs repeatedly but not continuously in the same place in the processor. As such, an intermittent fault manifests itself via intermittent errors.

The classification of faults and errors based on duration serves a useful purpose. The approach to tolerating a fault depends on its duration. Tolerating a permanent fault requires the ability to avoid using the faulty component, perhaps by using a fault-free replica of that component. Tolerating a transient fault requires no such self-repair because the fault will not persist. Fault tolerance schemes tend to treat intermittent faults as either transients or permanents, depending on how often they recur, although there are a few schemes designed specifically for tolerating intermittent faults [48].

1.2.3 Underlying Physical Phenomena

There are many physical phenomena that lead to faults, and we discuss them now based on their duration. Where applicable, we discuss techniques for reducing the likelihood of these physical phenomena leading to faults. Fault avoidance techniques are complementary to fault tolerance.

Transient phenomena. There are two well-studied causes of transient faults, and we refer the interested reader to the insightful historical study by Ziegler et al. [50] of IBM's experiences with soft errors. The first cause is cosmic radiation [49]. The cosmic rays themselves are not the culprits but rather the high-energy particles that are produced when cosmic rays impact the atmosphere. A computer can theoretically be shielded from these high-energy particles (at an extreme, by placing the computer in a cave), but such shielding is generally impractical. The second source of transient faults is alpha particles [22], which are produced by the natural decay of radioactive isotopes. The source of these radioactive isotopes is often, ironically, metal in the chip packaging itself. If a high-energy cosmic ray-generated particle or alpha particle strikes a chip, it can dislodge a significant amount of charge (electrons and holes) within the semiconductor material. If this charge exceeds the critical charge, often denoted Q_{crit}, of an SRAM or DRAM cell or p–n junction, it can flip the value of that cell or transistor output. Because the disruption is a one-time, transient event, the error will disappear once the cell or transistor's output is overwritten.

Transient faults can occur for reasons other than the two best-known causes described above. One possible source of transient faults is electromagnetic interference (EMI) from outside sources. A chip can also create its own EMI, which is often referred to as "cross-talk." Another source of transient errors is supply voltage droops due to large, quick changes in current draw. This source of errors is often referred to as the "dI/dt problem" because it depends on the current changing (dI) in a short amount of time (dt). Architects have recently explored techniques for reducing dI/dt, such as by managing the activity of the processor to avoid large changes in activity [26].

Permanent phenomena. Sources of permanent faults can be placed into three categories.

1. Physical wear-out: A processor in the field can fail because of any of several physical wear-out phenomena. A wire can wear out because of electromigration [7, 13, 18, 19]. A transistor's gate oxide can break down over time [6, 10, 21, 24, 29, 41]. Other physical phenomena that lead to permanent wear-outs include thermal cycling and mechanical stress. Many of these wear-out phenomena are exacerbated by increases in temperature. The RAMP model of Srinivasan et al. [40] provides an excellent tutorial on these four phenomena and a model for predicting their impacts on future technologies. The dependence of wear-out on temperature is clearly illustrated in the equations of the RAMP model.

    There has recently been a surge of research in techniques for avoiding wear-out faults. The group that developed the RAMP model [40] proposed the idea of lifetime reliability management [39]. The key insight of this work is that a processor can manage itself to achieve a lifetime reliability goal. A processor can use the RAMP model to estimate its expected lifetime and adjust itself—for example, by reducing its voltage and frequency—to either extend its lifetime (at the expense of performance) or improve its performance (at the expense of lifetime reliability). Subsequent research has proposed avoiding wear-out faults by using voltage and frequency scaling [42], adaptive body biasing [44], and by scheduling tasks on cores in a wear-out-aware fashion [12, 42, 44]. Other research has proposed techniques to avoid specific wear-out phenomena, such as negative bias temperature instability [1, 34]. More generally, dynamic temperature management [37] can help to alleviate the impact of wear-out phenomena that are exacerbated by increasing temperatures.

2. Fabrication defects: The fabrication of chips is an imperfect process, and chips may be manufactured with inherent defects. These defects may be detected by post-fabrication, pre-shipment testing, in which case the defect-induced faults are avoided in the field. However, defects may not reveal themselves until the chip is in the field. One particular concern for post-fabrication testing is that increasing leakage currents are making I_{DDQ} and burn-in testing infeasible [5, 31].

For the purposes of designing a fault tolerance scheme, fabrication defects are identical to wear-out faults, except that (a) they occur at time zero and (b) they are much more likely to occur "simultaneously"—that is, having multiple fabrication defects in a single chip is far more likely than having multiple wear-out faults occur at the same instant in the field.

3. Design bugs: Because of design bugs, even a perfectly fabricated chip may not behave correctly in all situations. Some readers may recall the infamous floating point division bug in the Intel Pentium processor [4], but it is by no means the only example of a bug in a shipped processor. Industrial validation teams try to uncover as many bugs as possible before fabrication, to avoid having these bugs manifest themselves as faults in the field, but the complete validation of a nontrivial processor is an intractable problem [3]. Despite expending vast resources on validation, there are still many bugs in recently shipped processors [2, 15–17]. Designing a scheme to tolerate design bugs poses some unique challenges, relative to other types of faults. Most notably, homogeneous spatial redundancy (e.g., TMR) is ineffective; all three replicas will produce the same erroneous result due to a design bug because the bug is present in all three replicas.

Intermittent phenomena. Some physical phenomena may lead to intermittent faults. The canonical example is a loose connection. As the chip temperature varies, a connection between two wires or devices may be more or less resistive and more closely model an open circuit or a fault-free connection, respectively. Recently, intermittent faults have been identified as an increasing threat largely due to temperature and voltage fluctuations, as well as prefailure component wear-out [8].

1.3 TRENDS LEADING TO INCREASED FAULT RATES

Fault-tolerant computer architecture has enjoyed a recent renaissance in response to several trends that are leading toward an increasing number of faults in commodity processors.

1.3.1 Smaller Devices and Hotter Chips

The dimensions of transistors and wires directly affect the likelihood of faults, both transient and permanent. Furthermore, device dimensions impact chip temperature, and temperature has a strong impact on the likelihood of permanent faults.

Transient faults. Smaller devices tend to have smaller critical charges, Q_{crit}, and we discussed in "Transient Phenomena" from Section 1.2.3 how decreasing Q_{crit} increases the probability that a high-energy particle strike can disrupt the charge on the device. Shivakumar et al. [35] analyzed the transient error trends for smaller transistors and showed that transient errors will become far more numerous in the future. In particular, they expect the transient error rate for combinational logic to increase dramatically and even overshadow the transient error rates for SRAM and DRAM.

Permanent faults. Smaller devices and wires are more susceptible to a variety of permanent faults, and this susceptibility is greatly exacerbated by process variability [5]. Fabrication using photolithography is an inherently imperfect process, and the dimensions of fabricated devices and wires may stray from their expected values. In previous generations of CMOS technology, this variability was mostly lost in the noise. A 2-nm variation around a 250-nm expected dimension is insignificant. However, as expected dimensions become smaller, variability's impact becomes more pronounced. A 2-nm variation around a 20-nm expected dimension can lead to a noticeable impact on behavior. Given smaller dimensions and greater process variability, there is an increasing likelihood of wires that are too small to support the required current density and transistor gate oxides that are too thin to withstand the voltages applied across them.

Another factor causing an increase in permanent faults is temperature. For a given chip area, trends are leading toward a greater number of transistors, and these transistors are consuming increasing amounts of active and static (leakage) power. This increase in power consumption per unit area translates into greater temperatures, and the RAMP model of Srinivasan et al. [40] highlights how increasing temperatures greatly exacerbate several physical phenomena that cause permanent faults. Furthermore, as the temperature increases, the leakage current increases, and this positive feedback loop with temperature and leakage current can have catastrophic consequences for a chip.

1.3.2 More Devices per Processor

Moore's law has provided architects with ever-increasing numbers of transistors per processor chip. With more transistors, as well as more wires connecting them, there are more opportunities for faults both in the field and during fabrication. Given even a constant fault rate for a single transistor, which is a highly optimistic and unrealistic assumption, the fault rate of a processor is increasing proportionately to the number of transistors per processor. Intuitively, the chances of one billion transistors all working correctly are far less than the probability of one million transistors all working correctly. This trend is unaffected by the move to multicore processors; it is the sheer number of devices per processor, not per core, that leads to more opportunities for faults.

1.3.3 More Complicated Designs

Processor designs have historically become increasingly complicated. Given an increasing number of transistors with which to work, architects have generally found innovative ways to modify microarchitectures to extract more performance. Cores, in particular, have benefitted from complex features such as dynamic scheduling (out-of-order execution), branch prediction, speculative load-store disambiguation, prefetching, and so on. An Intel Pentium 4 core is far more complicated than the original Pentium. This trend may be easing or even reversing itself somewhat because of power limitations—for example, Sun Microsystems' UltraSPARC T1 and T2 processors consist of numerous simple, in-order cores—but even processors with simple cores are likely to require complicated memory systems and interconnection networks to provide the cores with sufficient instruction and data bandwidth.

The result of increased processor complexity is a greater likelihood of design bugs eluding the validation process and escaping into the field. As discussed in "Permanent Phenomena" from Section 1.2.3 design bugs manifest themselves as permanent, albeit rarely exercised, faults. Thus, increasing design complexity is another contributor to increasing fault rates.

1.4 ERROR MODELS

Architects must be aware of the different types of faults that can occur, and they should understand the trends that are leading to increasing numbers of faults. However, architects rarely need to consider specific faults when they design processors. Intuitively, architects care about the possible errors that may occur, not the underlying physical phenomena. For example, an architect might design a cache frame such that it tolerates a single bit-flip error in the frame, but the architect's fault tolerance scheme is unlikely to be affected by which faults could cause a single bit-flip error.

Rather than explicitly consider every possible fault and how they could manifest themselves as errors, architects generally use *error models*. An error model is a simple, tractable tool for analyzing a system's fault tolerance. An example of an error model is the well-known "stuck-at" model, which models the impact of faults that cause a circuit value to be stuck at either 0 or 1. There are many underlying physical phenomena that can be represented with the stuck-at model, including some short and open circuits. The benefit of using an error model, such as the stuck-at model, instead of considering the possible physical phenomena, is that architects can design systems to tolerate errors within a set of error models. One challenge with error modeling, as with all modeling, is the issue of "garbage in, garbage out." If the error model is not representative of the errors that are likely to occur, then designing systems to tolerate these errors is not useful. For example, if we assume a stuck-at model for bits in a cache frame but an underlying physical fault causes a bit to instead take on the value of a neighboring bit, then our fault tolerance scheme may be ineffective.

There are many different error models, and we can classify them along three axes: type of error, error duration, and number of simultaneous errors.

1.4.1 Error Type

The stuck-at model is perhaps the best-known error model for two reasons. First, it represents a wide range of physical faults. Second, it is easy to understand and use. An architect can easily enumerate all possible stuck-at errors and analyze how well a fault tolerance scheme handles every possible error.

However, the stuck-at model does not represent the effects of many physical phenomena and thus cannot be used in all situations. If an architect uses the stuck-at error model when developing a fault tolerance scheme, then faults that do not manifest themselves as stuck-at errors may not be tolerated. If these faults are likely, then the system will be unreliable. Thus, other error models have been developed to represent the different erroneous behaviors that would result from underlying physical faults that do not manifest themselves as stuck-at errors.

One low-level error model, similar to stuck-at errors, is *bridging errors* (also known as *coupling errors*). Bridging errors model situations in which a given circuit value is bridged or coupled to another circuit value. This error model corresponds to many short-circuit and cross-talk fault scenarios. For example, the bridging error model is appropriate for capturing the behavior of a fabrication defect that causes a short circuit between two wires.

A higher-level error model is the *fail-stop error* model. Fail-stop errors model situations in which a component, such as a processor core or network switch, ceases to perform any function. This error model represents the impact of a wide variety of catastrophic faults. For example, chipkill memory [9, 14] is designed to tolerate fail-stop errors in DRAM chips regardless of the underlying physical fault that leads to the fail-stop behavior.

A relatively new error model is the *delay error* model, which models scenarios in which a circuit or component produces the correct value but at a time that is later than expected. Many underlying physical phenomena manifest themselves as delay errors, including progressive wear-out of transistors and the impact of process variability. Recent research called Razor [11] proposes a scheme for tolerating faults that manifest themselves as delay errors.

1.4.2   Error Duration

Error models have durations that are almost always classified into the same three categories described in Section 1.2.2: transient, intermittent, and permanent. For example, an architect could consider all possible transient stuck-at errors as his or her error model.

1.4.3   Number of Simultaneous Errors

A critical aspect of an error model is how many simultaneous errors it allows. Because physical faults have typically been relatively rare events, most error models consider only a single error at a time. To refine our example from the previous section, an architect could consider all possible *single* stuck-at errors as his or her error model. The possibility of multiple simultaneous errors is so unlikely that architects rarely choose to expend resources trying to tolerate these situations. Multiple-error scenarios are not only rare, but they are also far more difficult to reason about. Often, error models that permit multiple errors force architects to consider "offsetting errors," in which the affects of one error are hidden from the error detection mechanism by another error. For example, consider a system with a parity bit that protects a word of data. If one error flips a bit in that word and another error causes the parity check circuitry to erroneously determine that the word passed the parity check, then the corrupted data word will not be detected.

There are three reasons to consider error models with multiple simultaneous errors. First, for mission-critical computers, even a vanishingly small probability of a multiple error must be considered. It is not acceptable for these computers to fail in the presence of even a highly unlikely event. Thus, these systems must be designed to tolerate these multiple-error scenarios, regardless of the associated cost. Second, as discussed in Section 1.3, there are trends leading to an increasing number of faults. At some fault rate, the probability of multiple errors becomes nonnegligible and worth expending resources to tolerate, even for non-mission-critical computers. Third, the possibility of *latent errors*, errors that occur but are undetected and linger in the system, can lead to subsequent multiple-error scenarios. The presence of a latent error (e.g., a bit flip in a data word that has not been accessed in a long time) can cause the next error to appear to be a multiple simultaneous error, even if the two errors occur far apart in time. This ability of latent errors to confound error models motivates architects to design systems that detect errors quickly before another error can occur and thus violate the commonly used single-error model.

1.5 FAULT TOLERANCE METRICS

In this book, we present a wide range of approaches to tolerating the faults described in the past two sections. To evaluate these fault tolerance solutions, architects devise experiments to either test hypotheses or compare their ideas to previous work. These experiments might involve prototype hardware, simulations, or analytical models.

After performing experiments, an architect would like to present his or her results using appropriate metrics. For performance, we use a variety of metrics such as instructions per cycle or transactions per minute. For fault tolerance, we have a wide variety of metrics from which to choose, and it is important to choose appropriate metrics. In this section, we present several metrics and discuss when they are appropriate.

1.5.1 Availability

The availability of a system at time t is the probability that the system is operating correctly at time t. For many computing applications, availability is an appropriate metric. We want to improve the availability of the processors in desktops, laptops, servers, cell phones, and many other devices. The units for availability are often the "number of nines." For example, we often refer to a system with 99.999% availability as having "five nines" of availability.

1.5.2 Reliability

The *reliability* of a system at time t is the probability that the system has been operating correctly from time zero until time t. Reliability is perhaps the best-known metric, and a well-known word, but it is rarely an appropriate metric for architects. Unless a system failure is catastrophic (e.g., avionics), reliability is a less useful metric than availability.

1.5.3 Mean Time to Failure

Mean time to failure (MTTF) is often an appropriate and useful metric. In general, we wish to extend a processor's MTTF, but we must remember that MTTF is a mean and that mean values do not fully represent probability distributions. Consider two processors, P_A and P_B, which have MTTF values of 10 and 12, respectively. At first glance, based on the MTTF metric, P_B appears preferable. However, if the *variance* of failures is much higher for P_B than for P_A, as illustrated in the example in Table 1.1, then P_B might suffer more failures in the first 3 years than P_A. If we expect our computer to have a useful lifetime of 3 years before obsolescence, then P_A is actually preferable despite its smaller MTTF. To address this limitation of MTTF, Ramachandran et al. [28] invented the nMTTF metric. If nMTTF equals a time t, for some value of n, then the probability of failure of a given processor is $n/100$.

1.5.4 Mean Time Between Failures

Mean time between failures (MTBF) is similar to MTTF, but it also considers the time to repair. MTBF is the MTTF plus the mean time to repair (MTTR). Availability is a function of MTBF, that is,

$$\text{Availability} = \frac{\text{MTTF}}{\text{MTBF}} = \frac{\text{MTTF}}{\text{MTTF} + \text{MTTR}}$$

1.5.5 Failures in Time

The failures in time (FIT) rate of a component or a system is the number of failures it incurs over one billion (10^9) hours, and it is inversely proportional to MTTF. This is a somewhat odd and arbitrary metric, but it has been commonly used in the fault tolerance community. One reason for its use is that FIT rates can be added in an intuitive fashion. For example, if a system consisting of two components, A and B, fails if either component fails, then the FIT rate of the system is the FIT rate of A plus the FIT rate of B. The "raw" FIT rate of a component—the FIT rate if we do not consider failures that are architecturally masked—is often less informative than the effective FIT

	P_A	P_B
TABLE 1.1: Failure distributions for four chips each of P_A and P_B.		
lifetime of chip 1	9	2
lifetime of chip 2	10	2
lifetime of chip 3	10	21
lifetime of chip 4	11	23
mean lifetime	10	12
standard deviation of lifetime	0.5	10

rate, which does consider such masking. We discuss how to scale the raw FIT rate next when we discuss vulnerability.

1.5.6   Architectural Vulnerability Factor

Architectural vulnerability factor [23] is a recently developed metric that provides insight into a structure's vulnerability to transient errors. The idea behind AVF is to classify microprocessor state as either required for architecturally correct execution (ACE state) or not (un-ACE state). For example, the program counter (PC) is almost always ACE state because a corruption of the PC almost always causes a deviation from ACE. The state of the branch predictor is always un-ACE because any state produced by a misprediction will not be architecturally visible; the processor will squash this state when it detects that the branch was mispredicted. Between these two extremes of always ACE and never ACE, there are many structures that have state that is ACE some fraction of the time. The AVF of a structure is computed as the average number of ACE bits in the structure in a given cycle divided by the total number of bits in the structure. Thus, if many ACE bits reside in a structure for a long time, that structure is highly vulnerable.

AVF can be used to scale a raw FIT rate into an effective FIT rate. The effective FIT rate of a component is its raw FIT rate multiplied by its AVF. As an extreme example, a branch predictor has an effective FIT rate of zero because all failures are architecturally masked. AVF analysis helps to identify which structures are most vulnerable to transient errors, and it helps an architect to derate how much a given structure affects a system's overall fault tolerance. Wang et al. [46] showed that AVF analysis may overestimate vulnerability in some instances and thus provides an architect with a conservative lower bound on reliability.

1.6 THE REST OF THIS BOOK

Fault tolerance consists of four aspects:

- Error detection (Chapter 2): A processor cannot tolerate a fault if it is unaware of it. Thus, error detection is the most important aspect of fault tolerance, and we devote the largest fraction of the book to this topic. Error detection can be performed at various granularities. For example, a localized error detection mechanism might check the correctness of an adder's output, whereas a global or *end-to-end* error detection mechanism [32] might check the correctness of an entire core.

- Error recovery (Chapter 3): When an error is detected, the processor must take action to mask its effects from the software. A key to error recovery is not making any state visible to the software until this state has been checked by the error detection mechanisms. A common approach to error recovery is for a processor to take periodic checkpoints of its architectural state and, upon error detection, reload into the processor's state a checkpoint taken before the error occurred.

- Fault diagnosis (Chapter 4): Diagnosis is the process of identifying the fault that caused an error. For transient faults, diagnosis is generally unnecessary because the processor is not going to take any action to repair the fault. However, for permanent faults, it is often desirable to determine that the fault is permanent and then to determine its location. Knowing the location of a permanent fault enables a self-repair scheme to deconfigure the faulty component. If an error detection mechanism is localized, then it also provides diagnosis, but an end-to-end error detection mechanism provides little insight into what caused the error. If diagnosis is desired in a processor that uses an end-to-end error detection mechanism, then the architect must add a diagnosis mechanism.

- Self-repair (Chapter 5): If a processor diagnoses a permanent fault, it is desirable to repair or reconfigure the processor. Self-repair may involve avoiding further use of the faulty component or reconfiguring the processor to use a spare component.

In this book, we devote one chapter to each of these aspects. Because fault-tolerant computer architecture is such a large field and we wish to keep this book focused, there are several related topics that we do not include in this book, including:

- Mechanisms for reducing vulnerability to faults: Based on AVF analysis, there has been a significant amount of research in designing processors such that they are less vulnerable to faults [47, 38]. This work is complementary to fault tolerance.

- Schemes for tolerating CMOS process variability: Process variability has recently become a significant concern [5], and there has been quite a bit of research in designing processors that tolerate its effects [20, 25, 30, 43]. If process variability manifests itself as a fault, then its impact is addressed in this book, but we do not address the situations in which process variability causes other unexpected but nonfaulty behaviors (e.g., performance degradation).

- Design validation and verification: Before fabricating and shipping chips, their designs are extensively validated to minimize the number of design bugs that escape into the field. Perfect validation would obviate the need to detect errors due to design bugs, but realistic processor designs cannot be completely validated [3].

- Fault-tolerant I/O, including disks and network controllers: This book focuses on processors and memory, but we cannot forget that there are other components in computer systems.

- Approaches for tolerating software bugs: In this book, we present techniques for tolerating hardware faults, but tolerating hardware faults provides no protection against buggy software.

We conclude in Chapter 6 with a discussion of what the future holds for fault-tolerant computer architecture. We discuss trends, challenges, and open problems in the field, as well as synergies between fault tolerance and other aspects of architecture.

1.7   REFERENCES

[1] J. Abella, X. Vera, and A. Gonzalez. Penelope: The NBTI-Aware Processor. In *Proceedings of the 40th Annual IEEE/ACM International Symposium on Microarchitecture*, pp. 85–96, Dec. 2007.

[2] Advanced Micro Devices. Revision Guide for AMD Athlon64 and AMD Opteron Processors. Publication 25759, Revision 3.59, Sept. 2006.

[3] R. M. Bentley. Validating the Pentium 4 Microprocessor. In *Proceedings of the International Conference on Dependable Systems and Networks*, pp. 493–498, July 2001. doi:10.1109/DSN.2001.941434

[4] M. Blum and H. Wasserman. Reflections on the Pentium Bug. *IEEE Transactions on Computers*, 45(4), pp. 385–393, Apr. 1996. doi:10.1109/12.494097

[5] S. Borkar. Designing Reliable Systems from Unreliable Components: The Challenges of Transistor Variability and Degradation. *IEEE Micro*, 25(6), pp. 10–16, Nov./Dec. 2005. doi:10.1109/MM.2005.110

[6]   J. R. Carter, S. Ozev, and D. J. Sorin. Circuit-Level Modeling for Concurrent Testing of Operational Defects due to Gate Oxide Breakdown. In *Proceedings of Design, Automation, and Test in Europe (DATE)*, pp. 300–305, Mar. 2005. doi:10.1109/DATE.2005.94

[7]   J. J. Clement. Electromigration Modeling for Integrated Circuit Interconnect Reliability Analysis. *IEEE Transactions on Device and Materials Reliability*, 1(1), pp. 33–42, Mar. 2001. doi:10.1109/7298.946458

[8]   C. Constantinescu. Trends and Challenges in VLSI Circuit Reliability. *IEEE Micro*, 23(4), July–Aug. 2003. doi:10.1109/MM.2003.1225959

[9]   T. J. Dell. A White Paper on the Benefits of Chipkill-Correct ECC for PC Server Main Memory. IBM Microelectronics Division Whitepaper, Nov. 1997.

[10]   D. J. Dumin. Oxide Reliability: A Summary of Silicon Oxide Wearout, Breakdown and Reliability. World Scientific Publications, 2002.

[11]   D. Ernst et al. Razor: A Low-Power Pipeline Based on Circuit-Level Timing Speculation. In *Proceedings of the 36th Annual IEEE/ACM International Symposium on Microarchitecture*, Dec. 2003. doi:10.1109/MICRO.2003.1253179

[12]   S. Feng, S. Gupta, and S. Mahlke. Olay: Combat the Signs of Aging with Introspective Reliability Management. In *Proceedings of the Workshop on Quality-Aware Design*, June 2008.

[13]   A. H. Fischer, A. von Glasow, S. Penka, and F. Ungar. Electromigration Failure Mechanism Studies on Copper Interconnects. In *Proceedings of the 2002 IEEE Interconnect Technology Conference*, pp. 139–141, 2002. doi:10.1109/IITC.2002.1014913

[14]   IBM. Enhancing IBM Netfinity Server Reliability: IBM Chipkill Memory. IBM Whitepaper, Feb. 1999.

[15]   IBM. IBM PowerPC 750FX and 750FL RISC Microprocessor Errata List DD2.X, version 1.3, Feb. 2006.

[16]   Intel Corporation. Intel Itanium Processor Specification Update. Order Number 249720-00, May 2003.

[17]   Intel Corporation. Intel Pentium 4 Processor Specification Update. Document Number 249199-065, June 2006.

[18]   S. Krumbein. Metallic Electromigration Phenomena. *IEEE Transactions on Components, Hybrids, and Manufacturing Technology*, 11(1), pp. 5–15, Mar. 1988. doi:10.1109/33.2957

[19]   P.-C. Li and T. K. Young. Electromigration: The Time Bomb in Deep-Submicron ICs. *IEEE Spectrum*, 33(9), pp. 75–78, Sept. 1996.

[20]   X. Liang and D. Brooks. Mitigating the Impact of Process Variations on Processor Register Files and Execution Units. In *Proceedings of the 39th Annual IEEE/ACM International Symposium on Microarchitecture*, Dec. 2006.

[21] B. P. Linder, J. H. Stathis, D. J. Frank, S. Lombardo, and A. Vayshenker. Growth and Scaling of Oxide Conduction After Breakdown. In *41st Annual IEEE International Reliability Physics Symposium Proceedings*, pp. 402–405, Mar. 2003. doi:10.1109/RELPHY.2003.1197781

[22] T. May and M. Woods. Alpha-Particle-Induced Soft Errors in Dynamic Memories. *IEEE Transactions on Electronic Devices*, 26(1), pp. 2–9, 1979.

[23] S. S. Mukherjee, C. Weaver, J. Emer, S. K. Reinhardt, and T. Austin. A Systematic Methodology to Compute the Architectural Vulnerability Factors for a High-Performance Microprocessor. In *Proceedings of the 36th Annual IEEE/ACM International Symposium on Microarchitecture*, Dec. 2003. doi:10.1109/MICRO.2003.1253181

[24] S. Oussalah and F. Nebel. On the Oxide Thickness Dependence of the Time-Dependent Dielectric Breakdown. In *Proceedings of the IEEE Electron Devices Meeting*, pp. 42–45, June 1999. doi:10.1109/HKEDM.1999.836404

[25] S. Ozdemir, D. Sinha, G. Memik, J. Adams, and H. Zhou. Yield-Aware Cache Architectures. In *Proceedings of the 39th Annual IEEE/ACM International Symposium on Microarchitecture*, pp. 15–25, Dec. 2006.

[26] M. D. Powell and T. N. Vijaykumar. Pipeline Damping: A Microarchitectural Technique to Reduce Inductive Noise in Supply Voltage. In *Proceedings of the 30th Annual International Symposium on Computer Architecture*, pp. 72–83, June 2003. doi:10.1109/ISCA.2003.1206990

[27] D. K. Pradhan. *Fault-Tolerant Computer System Design*. Prentice-Hall, Inc., Upper Saddle River, NJ, 1996.

[28] P. Ramachandran, S. V. Adve, P. Bose, and J. A. Rivers. Metrics for Architecture-Level Lifetime Reliability Analysis. In *Proceedings of the International Symposium on Performance Analysis of Systems and Software*, pp. 202–212, Apr. 2008.

[29] R. Rodriguez, J. H. Stathis, and B. P. Linder. Modeling and Experimental Verification of the Effect of Gate Oxide Breakdown on CMOS Inverters. In *Proceedings of the IEEE International Reliability Physics Symposium*, pp. 11–16, 2003. doi:10.1109/RELPHY.2003.1197713

[30] B. F. Romanescu, M. E. Bauer, D. J. Sorin, and S. Ozev. Reducing the Impact of Intra-Core Process Variability with Criticality-Based Resource Allocation and Prefetching. In *Proceedings of the ACM International Conference on Computing Frontiers*, pp. 129–138, May 2008. doi:10.1145/1366230.1366257

[31] S. S. Sabade and D. Walker. IDDQ Test: Will It Survive the DSM Challenge? *IEEE Design & Test of Computers*, 19(5), pp. 8–16, Sept./Oct. 2002.

[32] J. H. Saltzer, D. P. Reed, and D. D. Clark. End-to-End Arguments in Systems Design. *ACM Transactions on Computer Systems*, 2(4), pp. 277–288, Nov. 1984. doi:10.1145/357401.357402

[33] O. Serlin. Fault-Tolerant Systems in Commercial Applications. *IEEE Computer*, 17(8), pp. 19–30, Aug. 1984.

[34]   J. Shin, V. Zyuban, P. Bose, and T. M. Pinkston. A Proactive Wearout Recovery Approach for Exploiting Microarchitectural Redundancy to Extend Cache SRAM Lifetime. In *Proceedings of the 35th Annual International Symposium on Computer Architecture*, pp. 353–362, June 2008. doi:10.1145/1394608.1382151

[35]   P. Shivakumar, M. Kistler, S. W. Keckler, D. Burger, and L. Alvisi. Modeling the Effect of Technology Trends on the Soft Error Rate of Combinational Logic. In *Proceedings of the International Conference on Dependable Systems and Networks*, June 2002. doi:10.1109/DSN.2002.1028924

[36]   D. P. Siewiorek and R. S. Swarz. *Reliable Computer Systems: Design and Evaluation.* A. K. Peters, third edition, Natick, Massachusetts, 1998.

[37]   K. Skadron, M. R. Stan, W. Huang, S. Velusamy, K. Sankaranarayanan, and D. Tarjan. Temperature-aware Microarchitecture. In *Proceedings of the 30th Annual International Symposium on Computer Architecture*, pp. 2–13, June 2003. doi:10.1145/859619.859620

[38]   N. Soundararajan, A. Parashar, and A. Sivasubramaniam. Mechanisms for Bounding Vulnerabilities of Processor Structures. In *Proceedings of the 34th Annual International Symposium on Computer Architecture*, pp. 506–515, June 2007. doi:10.1145/1250662.1250725

[39]   J. Srinivasan, S. V. Adve, P. Bose, and J. A. Rivers. The Case for Lifetime Reliability-Aware Microprocessors. In *Proceedings of the 31st Annual International Symposium on Computer Architecture*, June 2004. doi:10.1109/ISCA.2004.1310781

[40]   J. Srinivasan, S. V. Adve, P. Bose, and J. A. Rivers. The Impact of Technology Scaling on Lifetime Reliability. In *Proceedings of the International Conference on Dependable Systems and Networks*, June 2004. doi:10.1109/DSN.2004.1311888

[41]   J. H. Stathis. Physical and Predictive Models of Ultrathin Oxide Reliability in CMOS Devices and Circuits. *IEEE Transactions on Device and Materials Reliability*, 1(1), pp. 43–59, Mar. 2001. doi:10.1109/7298.946459

[42]   D. Sylvester, D. Blaauw, and E. Karl. ElastIC: An Adaptive Self-Healing Architecture for Unpredictable Silicon. *IEEE Design & Test of Computers*, 23(6), pp. 484–490, Nov./Dec. 2006.

[43]   A. Tiwari, S. R. Sarangi, and J. Torrellas. ReCycle: Pipeline Adaptation to Tolerate Process Variability. In *Proceedings of the 34th Annual International Symposium on Computer Architecture*, June 2007.

[44]   A. Tiwari and J. Torrellas. Facelift: Hiding and Slowing Down Aging in Multicores. In *Proceedings of the 41st Annual IEEE/ACM International Symposium on Microarchitecture*, pp. 129–140, Nov. 2008.

[45]   J. von Neumann. Probabilistic Logics and the Synthesis of Reliable Organisms from Unreliable Components. In C. E. Shannon and J. McCarthy, editors, *Automata Studies*, pp. 43–98. Princeton University Press, Princeton, NJ, 1956.

[46] N. J. Wang, A. Mahesri, and S. J. Patel. Examining ACE Analysis Reliability Estimates Using Fault-Injection. In *Proceedings of the 34th Annual International Symposium on Computer Architecture*, June 2007. doi:10.1145/1250662.1250719

[47] C. Weaver, J. Emer, S. S. Mukherjee, and S. K. Reinhardt. Techniques to Reduce the Soft Error Rate of a High-Performance Microprocessor. In *Proceedings of the 31st Annual International Symposium on Computer Architecture*, pp. 264–275, June 2004. doi:10.1109/ISCA.2004.1310780

[48] P. M. Wells, K. Chakraborty, and G. S. Sohi. Adapting to Intermittent Faults in Multicore Systems. In *Proceedings of the Thirteenth International Conference on Architectural Support for Programming Languages and Operating Systems*, Mar. 2008. doi:10.1145/1346281.1346314

[49] J. Ziegler. Terrestrial Cosmic Rays. *IBM Journal of Research and Development*, 40(1), pp. 19–39, Jan. 1996.

[50] J. Ziegler et al. IBM Experiments in Soft Fails in Computer Electronics. *IBM Journal of Research and Development*, 40(1), pp. 3–18, Jan. 1996.

CHAPTER 2

Error Detection

Error detection is the most important aspect of fault tolerance because a processor cannot tolerate a problem of which it is not aware. Even if the processor cannot recover from a detected error, the processor can still alert the user that an error has occurred and halt. Error detection thus provides, at the minimum, a measure of *safety*. A safe processor does not do anything incorrect. Without recovery, the processor may not be able to make forward progress, but at least it is safe. It is far preferable for a processor to do nothing than to silently fail and corrupt data.

In this chapter, as well as subsequent chapters, we divide our discussion into general concepts and domain-specific solutions. These processor domains include microprocessor cores (Section 2.2), caches and memories (Section 2.3), and multicore memory systems (Section 2.4). We divide the discussion in this fashion because the issues in each domain tend to be quite distinct.

2.1 GENERAL CONCEPTS

There are some fundamental concepts in error detection that we discuss now, so as to better understand the applications of these concepts to specific domains. The key to error detection is redundancy: a processor with no redundancy fundamentally cannot detect any errors. The question is not whether to use redundancy but rather what kind of redundancy should be used. The three classes of redundancy—physical (sometimes referred to as "spatial"), temporal, and information—are described in Table 2.1. All error detection schemes use one or more of these types of redundancy, and we now discuss each in more depth.

2.1.1 Physical Redundancy

Physical (or spatial) redundancy is a commonly used approach for providing error detection. The simplest form of physical redundancy is dual modular redundancy (DMR) with a comparator, illustrated in Figure 2.1. DMR provides excellent error detection because it detects all errors except for errors due to design bugs, errors in the comparator, and unlikely combinations of simultaneous errors that just so happen to cause both modules to produce the same incorrect outputs.

Adding an additional replica and replacing the comparator with a voter leads to the classic triple modular redundant design, shown in Figure 2.2. With triple modular redundancy (TMR),

TABLE 2.1: The Three Types of Redundancy.		
TYPE OF REDUNDANCY	**BASIC IDEA**	**SINGLE EXAMPLE**
Physical (spatial)	Add redundant hardware	Replicate a module and have the two replicas compare their results
Temporal	Perform redundant operations	Run a program twice on the same hardware and compare the results of the two executions
Information	Add redundant bits to a datum	Add a parity bit to a word in memory

the output of the majority of the modules is chosen by the voter to be the output of the system. TMR offers error detection that is comparable to DMR. TMR's advantage is that, for single errors, it also provides fault diagnosis (the outvoted module has the fault) and error recovery (the system continues to run in the presence of the error). A more general physical redundancy scheme is N-modular redundancy (NMR) [86], which, for odd values of N greater than three, provides better error detection coverage, diagnosis, and recovery than TMR.

Physical redundancy can be implemented at various granularities. At a coarse grain, we can replicate an entire processor or replicate cores within a multicore processor. At a finer grain, we can replicate an ALU or a register. Finer granularity provides finer diagnosis, but it also increases the relative overhead of the voter. Taken to an absurdly fine extreme, using TMR at the granularity of a single NAND gate would create a scenario in which the voter was larger than the three modules.

Physical redundancy does not have to be homogeneous. That is, the redundant hardware does not have to be identical to the original hardware. Heterogeneity, also called "design diversity" [6], can serve two purposes.

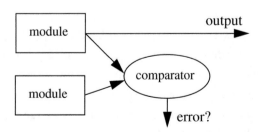

FIGURE 2.1: Dual modular redundancy.

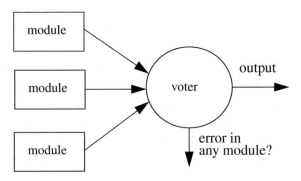

FIGURE 2.2: Triple modular redundancy.

First, it enables detection of errors due to design bugs. The Boeing 777 [93] uses heteroge-neous "triple-triple" modular redundancy, as illustrated in Figure 2.3. This design uses heteroge-neous processors within each unit and thus a design bug in any of the processors will be detected (and corrected) by the other two processors in the unit. The second benefit of heterogeneity is the ability to reduce the cost of the redundant hardware, as compared to homogeneous redundancy. In many situations, it is easier to check that an operation is performed correctly than to perform the operation; in these situations, a heterogeneous checker can be smaller and cheaper than the unit it is checking. An extreme example of heterogeneous hardware redundancy is a *watchdog timer* [42]. A watchdog timer is a piece of hardware that monitors other hardware for signs of liveness. For exam-ple, a processor's watchdog timer might track memory requests on the bus. If no requests have been observed for an extremely long time that exceeds a predefined threshold, then the watchdog timer reports that an error has occurred. Checking a processor's liveness is far simpler than performing

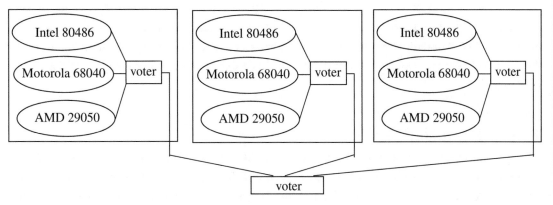

FIGURE 2.3: Boeing 777's triple TMR [93].

all of the processor's operations, and a watchdog timer can thus be far cheaper than a redundant processor.

The primary costs of physical redundancy are the hardware cost and power and energy consumption. For example, compared to an unprotected system, a system with TMR uses more than three times as much hardware (two redundant modules and a voter) and a corresponding extra amount of power and energy. For mission-critical systems that require the error detection capability of NMR, these costs may be unavoidable, but these costs are rarely acceptable for commodity processors. In particular, as modern processors try to extract as much performance as possible for a given energy and power budget, NMR's power and energy costs are almost certainly impractical. Also, when using NMR, a designer must remember that N times as much hardware is susceptible to N times as many errors, if we assume a constant error rate per unit of hardware.

2.1.2 Temporal Redundancy

In its most basic form, temporal redundancy requires a unit to perform an operation twice (or more times, in theory, but we only consider two iterations here), one after the other, and then compare the results. Thus, the total time is doubled, ignoring the latency to compare the results, and the performance of the unit is halved. Unlike with physical redundancy, there is no extra hardware or power cost (once again ignoring the comparator). However, as with DMR, the active energy consumption is doubled because twice as much work is performed.

Because of temporal redundancy's steep performance cost, many schemes use pipelining to hide the latency of the redundant operation. As one example, consider a fully pipelined unit, such as a multiplier. Assume that a multiplication takes X cycles to complete. If we begin the initial computation on cycle C, we can begin the redundant computation on cycle $C+1$. The latency of the checked multiplication is only increased by one cycle; instead of completing on cycle $C+X$, it now completes on cycle $C+X+1$. This form of temporal redundancy reduces the latency penalty significantly, but it still has a throughput penalty because the multiplier can perform only half as many unique (nonredundant) multiplications per unit of time. This form of temporal redundancy does not address the energy penalty at all; it still uses twice as much active energy as a nonredundant unit.

2.1.3 Information Redundancy

The basic idea behind information redundancy is to add redundant bits to a datum to detect when it has been affected by an error. An *error-detecting code* (*EDC*) maps a set of 2^k k-bit datawords to a set of 2^k n-bit "codewords," where $n > k$. The key idea is to map the datawords to codewords such that the codewords are as "far apart" from each other as possible in the n-dimensional codeword space.

The distance between any two codewords, called the *Hamming distance* (HD), is the number of bit positions in which they differ. For example, 01110 and 11010 differ in two bit positions.

The HD of an EDC is the minimum HD between any two codewords, and the EDC's HD is what determines how many single bit-flip errors it can detect in a single codeword. The two examples in Figure 2.4 pictorially illustrate two EDCs, one with an HD of two and the other with three. In the HD=2 example, we observe that, for any legal codeword, an error in any one of its bits will transform the codeword into an illegal word in the codeword space. For example, a single-bit error might transform 011 into 111, 001, or 010; none of these three words is a legal codeword. Thus, a single-bit error will always be detected because it will lead to an illegal word. A double-bit error might transform 011 into 000, which is also a legal codeword and would thus be undetected. In the HD=3 example, for either legal codeword, an error in any one or two of its bits will transform the codeword into an illegal word. Thus, a single-bit or double-bit error will always be detected. More generally, an EDC can detect errors in up to HD–1 bit positions.

The simplest and most common EDC is *parity*. Parity adds one *parity bit* to a dataword to convert it into a codeword. For even (odd) parity, the parity bit is added such that the total number of ones in the codeword is even (odd). Parity is an HD=2 EDC that can thus detect single-bit errors. Parity is popular because it is simple and inexpensive to implement, and it provides decent error detection coverage.

More sophisticated codes with larger HDs can detect more errors, and many of these codes can also *correct* errors. An *error-correcting code* (*ECC*) adds enough redundant bits to provide correction. For example, the HD=3 code in Figure 2.4 can correct single-bit errors. Consider the three possible single-bit errors in the codeword 000: 001, 010, and 100. All three of these codewords are

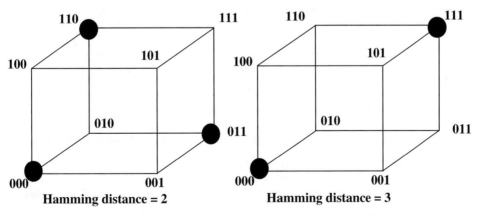

FIGURE 2.4: Hamming distance examples. Black circles denote legal codewords. Vertices without black circles correspond to illegal words in the codeword space.

closer to 000 than they are to the next nearest codeword, 111. Thus, the code would correct the error by interpreting 001, 010, or 100 as being the codeword 000. An ECC can correct errors in up to (HD−1)/2 bit positions. In Figure 2.5, we illustrate a more efficient HD=3 ECC known as a *Hamming (7,4)* code because codewords are 7 bits and datawords are 4 bits. This ECC, like the simpler but less efficient HD=3 code in Figure 2.4, can also correct a single-bit error. The Hamming (7,4) code has an overhead that is 3 bits per 4-bit dataword compared to the simpler code that adds 2 bits per 1-bit dataword.

Error codes are often classified based on their detection and correction abilities. A common classification is SECDED, which stands for "single-error correcting (SEC) and double-error detecting (DED)" and has an HD of 4. Note that the HD=3 example in Figure 2.4 can either correct single errors *or* detect single or double errors, but it cannot do both. For example, if this code is to

---

Creating a codeword. Given a 4-bit dataword $\underline{D}$ = [d1 d2 d3 d4], we construct a 7-bit codeword $\underline{C}$ by computing three overlapping parity bits:

p1 = d1 xor d2 xor d4

p2 = d1 xor d3 xor d4

p4 = d2 xor d3 xor d4

The 7-bit codeword $\underline{C}$ = [p1 p2 d1 p4 d2 d3 d4].

Correcting errors in a possibly corrupted codeword. Given a 7-bit word $\underline{R}$, we check it by multiplying it with the parity check matrix matrix $\underline{H}$ below:

p1	p2	d1	p4	d2	d3	d4
1	0	1	0	1	0	1
0	1	1	0	0	1	1
0	0	0	1	1	1	1

If $\underline{R}$ is a valid codeword, then $\underline{HR}$=$\underline{0}$, and no error correction is required.

Else, if $\underline{R}$ is a corrupted codeword, then $\underline{HR}$=$\underline{S}$, where the 3-bit $\underline{S}$ indicates the error's location.

Example 1: $\underline{R}$ = [0100101]. $\underline{HC}$ = [0 0 0] = $\underline{0}$ --> no error

Example 2: $\underline{R}$ = [01**1**0101] (error in bit position 3). $\underline{HC}$ = [1 1 0] --> we read the syndrome backwards to determine that the error location is in bit position 011 = 3

FIGURE 2.5: Hamming (7,4) code.

be used for SEC instead of DED, then a 001 would be corrected to be 000 instead of considering the possibility that a double-error had turned a 111 into 001. SECDED codes are commonly used for a variety of dataword sizes.

In Table 2.2, we show the relationship between dataword size and codeword size, for dataword sizes ranging from 8 to 256 bits.

We summarize the error detection and correction capabilities of error codes in Table 2.3. In this table, we include the capability to correct *erasures*. An erasure is a bit that is unreadable; the logic cannot tell if it is a 0 or a 1. Erasures are common in network communications, and they also occur in storage structures when a portion of the storage (e.g., a DRAM chip or a disk in a RAID array) is unresponsive because of a catastrophic failure. Correcting an erasure is easier than correcting an error because, with an erasure, we know the location of the erased bit. For example, consider an 8-bit dataword with a single parity bit. This parity bit can be used to detect a single error or to correct a single erasure, but it is insufficient to correct a single error.

There exist many error codes, and discussing them in depth is beyond the scope of this book. For a more complete treatment of the topic, we refer the interested reader to Wakerly's [88] excellent book on EDCs.

2.1.4   The End-to-End Argument

We can apply redundancy to detect errors at many different levels in the system—at the transistor, gate, cache block, core, and so on. A question for a computer architect is what level or levels are appropriate. Saltzer et al. [64] argued for "end-to-end" error detection in which we strive to

TABLE 2.2: SECDED Codes for Various Dataword Sizes.		
DATAWORD SIZE (BITS)	MINIMUM CODEWORD SIZE (BITS)	SECDED STORAGE OVERHEAD (%)
8	13	62.5
16	22	37.5
32	39	21.9
64	72	12.5
128	137	7.0
256	266	3.9

ERRORS DETECTED	ERRORS CORRECTED	ERASURES CORRECTED	MINIMUM HAMMING DISTANCE
D	0	0	D+1
0	0	E	E+1
0	C	0	2C+1
D	C	0	2C+D+1
D	0	E	D+E+1
0	C	E	2C+E+1
D	C	E	2C+D+E+1

TABLE 2.3: Summary of EDC and ECC Capabilities.

perform error detection at the "ends" or the highest level possible. Instead of adding hardware to immediately detect errors as soon as they occur, the end-to-end argument suggests that we should wait to detect errors until they manifest themselves as anomalous higher-level behaviors. For example, instead of detecting that a bit flipped, we would prefer to wait until that bit flip resulted in an erroneous instruction result or a program crash. By checking at a higher level, we can reduce the hardware costs and reduce the number of false positives (detected errors that have no impact on the core's behavior). Furthermore, we *have* to check at the ends anyway because only at the ends does the system have sufficient semantic knowledge to detect certain types of errors.

Relying only on end-to-end error detection has three primary drawbacks. First, detecting a high-level error like a program crash provides little diagnostic information. If the crash is due to a permanent fault, it would be beneficial to have some idea of where the fault is that caused the crash, or even that the crash was due to a physical fault and not a software bug. If only end-to-end error detection is used, then additional diagnostic mechanisms may be necessary.

The second drawback to relying only on high-level error detection is that it has a longer— and potentially unbounded—error detection latency. A low-level error like a bit flip may not result in a program crash for a long time. A longer error detection latency poses two challenges. First, to recover from a crash requires the processor to recover to a state from before the error's occurrence. Longer detection latencies thus require the processor to keep saved recovery points from further in the past. Unbounded detection latencies imply that certain detected errors will be unrecoverable because no prefault recovery point will exist. Second, longer detection latency means that the effects of an error may propagate farther. To avoid having an error propagate to the "outside world"—that is, a component outside what the core can recover in the case an error is detected, such as a printer or a network—the core must refrain from sending data to the outside world until it has been checked for errors. This fundamental issue in fault tolerance is called the *output commit problem* [26]. A longer detection latency exacerbates the output commit problem and leads to longer latencies for communicating data to the outside world.

The third drawback of relying solely on end-to-end error detection is that the recovery process itself may be more complicated. Recovering the state of a small component is often easier than recovering a larger component or an entire system. For example, consider a multicore processor. Recovering a single core is far easier than recovering the entire multicore processor. As we will explain in Chapter 3, recovering a multicore requires recovery of the state of the communication between the cores. As another example, IBM moved from a z9 processor design in which recovery was performed on a pair of lockstepped cores to a z10 processor design in which recovery is performed within a core [19]. One rationale for this design change was the complexity of recovering pairs of cores.

Because of both the benefits and drawbacks of end-to-end error detection, many systems use a combination of end-to-end and localized detection mechanisms. For example, networks often use both link-level (localized) retry and end-to-end checksums.

2.2 MICROPROCESSOR CORES

Having discussed error detection in general, we now discuss how this redundancy is applied in practice within microprocessor cores. We begin with functional unit and register file checking and then present a wide variety of more comprehensive error detection schemes.

2.2.1 Functional Units

There is a long history of error detection for functional units, and Sellers et al. [69] presented an excellent survey of checkers for functional units of all kinds. We refer the interested reader to the book by Sellers et al. for an in-depth treatment of this topic. In this section, we first discuss some general techniques before briefly discussing checkers that are specific to adders and multipliers because these are common functional units with well-studied solutions for error detection.

General Techniques. To detect errors in a functional unit, we could simply treat the unit as a black box and use physical or temporal redundancy. However, because we know something about the unit, we can develop error detection schemes that are more efficient. In particular, we can leverage knowledge of the mathematical operation performed by the functional unit.

One general approach to functional unit error detection is to use *arithmetic codes*. An arithmetic code is a type of EDC that is preserved by the functional unit. If a functional unit operates on input operands that are codewords in an arithmetic code, then the result of an error-free operation will also be a codeword. A functional unit is *fault-secure* if, for every possible fault in the fault model, there is no combination of valid codeword inputs that results in a codeword output.

A simple example of an arithmetic code that is preserved across addition is a code that takes an integer data word and multiplies it by an integer (e.g., 10). Assume we wish to add $A+B=C$. If we add $10A+10B$, we get $10C$ in the error-free case. However, if the error causes the adder to produce a result that is not a multiple of 10, then an error is detected. More sophisticated arithmetic codes rely on properties such as the relationship between the number of ones in the input codewords and the number of ones in the output codeword. Despite their great potential to detect errors in functional units, arithmetic codes are currently rarely used in commodity cores because of the large cost for the additional circuitry and the latencies to convert between datawords and codewords.

Another approach to functional unit error detection is a variant of temporal redundancy that can detect errors due to permanent faults. A permanently faulty functional unit that is protected with pure temporal redundancy computes the same incorrect answer every time it operates on the same operands; the redundant computations are equal and thus the errors are undetected. Reexecution with shifted operands (RESO) [56] overcomes this limitation by shifting the input operands before the redundant computation. The example in Figure 2.6 illustrates how RESO detects an error due to a permanent fault in an adder. Note that a RESO scheme that shifts by k bits requires an adder that is k-bits wider than normal.

Adders. Because adders are such fundamental components of all cores, there has been a large amount of research in detecting errors in them. Nicolaidis presents self-checking versions of several types of adders, including carry look-ahead [53]. Townsend et al. [83] developed a self-checking and self-correcting adder that combines TMR and temporal redundancy. There are also many error

Original Addition	Shifted-left-by-2 Addition
X X 0 0 1 0	0 0 1 0 X X
+ X X 1 0 0 1	+ 1 0 0 1 X X
X X 1 0 1 0	1 0 1 1 X X
↗ erroneous bit	↗ correct bit

FIGURE 2.6: Example of RESO. By comparing output bit 0 of the original addition to output bit 2 of the shifted-left-by-2 addition, RESO detects an error in the ALU. If this error was due to a permanent fault, it would not be detected by normal (nonshifted) reexecution because the results of the original and reexecuted addition would be equal.

detection schemes that apply to only specific types of adders. For example, there are self-checking techniques for carry-look-ahead adders [38] and carry select adders [70, 78, 85].

Multipliers. An efficient way to detect errors in multiplication is to use a modulo (or "residue") checking scheme. The key to modulo checking is that: $A \times B = C \rightarrow [(A \bmod M) \times (B \bmod M)] \bmod M = C \bmod M$. Thus, we can check the multiplication of $A \times B$ by checking if $[(A \bmod M) \times (B \bmod M)]$ $\bmod M = C \bmod M$. This result is interesting because, with an appropriate choice of M, the modulus operation can be performed with little hardware and the multiplication of $(A \bmod M)$ and $(B \bmod M)$ requires a far smaller multiplier than that required to multiply A and B. The total hardware for the checker is far smaller than the original multiplier. The only drawback to modulo checking is the probability of *aliasing*. That is, there is a nonzero probability that the multiplier erroneously computes $A \times B = D$ (where D does not equal C), but $[(A \bmod M) \times (B \bmod M)] \bmod M = D \bmod M$. This probability can be made arbitrarily small, but nonzero, through the choice of M. As M becomes smaller, the probability of aliasing increases. This result is intuitive because a smaller value of M means that we are hashing the operands and results into shorter lengths that have fewer unique values.

2.2.2   Register Files

A core's register file holds a significant amount of architectural state that must be kept error-free. As with any kind of storage structure, a simple approach to detecting errors is to use EDC or ECC. To reduce the storage and performance overheads of error codes, there has been some recent research to selectively protect only those registers that are predicted to be most vulnerable to faults. Intuitively, not all registers hold live values, and protecting dead values is unnecessary.

Blome et al. [9] developed a register value cache (RVC) that holds replicas of live register values. When the core wishes to read a register, it reads from both the original register file and the RVC. If the read hits in the RVC, then the two values are compared. If they are unequal, an error has been detected. Similarly, Montesinos et al. [49] realized that protecting all registers is unnecessary, and they proposed maintaining ECC only for those registers predicted to be most vulnerable to soft errors.

2.2.3   Tightly Lockstepped Redundant Cores

A straightforward application of physical redundancy is to simply replicate a core and create either a DMR or TMR configuration. The cores operate in tight lockstep and compare their results after every instruction or perhaps less frequently. The frequency of comparison determines the maximum error detection latency. This conceptually simple design has the benefits and drawbacks explained in Section 2.1.1. Because of its steep costs, it has traditionally been used only in highly reliable systems—like mainframes [73], the Tandem S2 [32], and the Hewlett Packard NonStop series up until the NonStop Advanced Architecture [7]—and mission-critical systems like the processor in the Boeing 777 [93].

With the advent of multicore processors and the difficulty of keeping all of these cores busy with useful work and fed with data from off-chip, core redundancy has become more appealing. The opportunity cost of using cores to run redundant threads may be low, although the power and energy costs are still significant. Exploiting these trends, recent work by Aggarwal et al. [3] described a multicore processor that uses DMR and TMR configurations for detecting and correcting errors. The dynamic core coupling (DCC) of LaFrieda et al. [36] shows how to dynamically, rather than statically, group the cores into DMR or TMR configurations.

2.2.4   Redundant Multithreading Without Lockstepping

Similar to the advent of multicore processors, the advent of simultaneously multithreaded (SMT) cores [84], such as the Intel Pentium 4 [12], provided an opportunity for low-cost redundancy. An SMT core with T thread contexts can execute T software threads at the same time. If an SMT core has fewer than T useful threads to run, then using otherwise idle thread contexts to run redundant threads provides cheap error detection. Redundant multithreading, depending on its implementation, may require little additional hardware beyond a comparator to determine whether the redundant threads are behaving identically. Redundant multithreading on an SMT core has less performance impact than a pure temporal redundancy scheme; its main impact on performance is because of the extra contention for core resources due to the redundant threads [77]. This contention can lead to queuing delays for the nonredundant threads. Redundant multithreading does have an opportunity cost, though, because thread contexts that run redundant threads are not available to run useful nonredundant work.

Rotenberg's paper on AR-SMT [62] was the first to introduce the idea of redundant multithreading on an SMT core. The active (A) and redundant (R) threads run simultaneously, but with a slight lag between them. The A-thread runs ahead of the R-thread and places the results of each committed instruction in a FIFO delay buffer. The R-thread compares the result of each instruction it completes with the A-thread's result in the delay buffer. If they are equal, the R-thread commits its instruction. Because the R-thread only commits instructions that have been successfully compared, its committed state is an error-free *recovery point*, that is, a point to which the processor may recover after detecting an error. Thus, if the R-thread detects that its instruction has a result different from that of the A-thread, it triggers an error and both threads recover to the most recently committed state of the R-thread.

When the delay buffer is full, the A-thread cannot complete more instructions; when the delay buffer is empty, the R-thread cannot commit more instructions. By allowing the A-thread to commit instructions before the comparison, AR-SMT avoids some performance penalties. Gomaa et al. [29] later showed that this design decision is particularly important when running the redundant threads on multiple cores because of the long latency to communicate results from the A-thread to the R-thread.

AR-SMT, as a temporal redundancy scheme, detects a wide range of transient errors. It may also detect some errors due to permanent faults if, by chance, one of the two threads (but not both) uses the permanently faulty hardware to execute an instruction. In an SMT core, this situation can occur, for example, with ALUs because there are multiple ALUs and there are no restrictions regarding which ALU each instruction will use. To extend redundant multithreading to consistently detect errors due to hard faults, the BlackJack technique [67] guarantees that the A-thread and R-thread will use different resources. The resources are coarsely divided into front-end and back-end pipeline resources to facilitate reasoning about what resources are used by which instructions. BlackJack is thus a combined temporal and physical redundancy scheme, although the physical redundancy is, in a sense, "free" because it already exists within the superscalar core.

AR-SMT inspired a large amount of work in redundant multithreading on both SMT cores and multicore processors. The goals of this subsequent work were to study implementations in greater depth and detail, reduce the performance impact, and reduce the hardware cost. Because there are so many papers in this area, we present only a few highlights here.

Options for Where and When to Compare Threads. Reinhardt and Mukherjee [60] developed a simultaneous and redundantly threaded (SRT) core that decreases the performance impact of AR-SMT by more carefully managing core resources and by more efficiently comparing the behaviors of the two threads. They also introduced the notion of "sphere of replication," which defines exactly which components are (and are not) protected by SRT. Explicitly considering the sphere of replication enables designers to more clearly reason about what needs to be replicated (e.g., is the thread replicated before or after each instruction is fetched?) and when comparisons need to occur (e.g., at every store or at every I/O event). For example, if the thread is replicated after each instruction is fetched, then the sphere of replication does not include the fetch logic and the scheme cannot detect errors in fetch. Similarly, if the redundant threads share a data cache and only the R-thread performs stores, after comparing its stores to those that the A-thread wishes to perform, then the data cache is outside the sphere of replication.

Smolens et al. [74] analyzed the tradeoffs between different spheres of replication. In particular, they studied how the point of comparison determines how much thread behavior history must be compared and the latency to detect errors. They then dramatically optimized the storage and comparison of thread histories by using a small amount of hardware to compute a fixed-length "fingerprint" or signature of each history. The threads' fingerprints are compared at the end of every checkpointing interval. Fingerprinting thus extends the error detection latency, compared to a scheme that compares the threads on a per-instruction basis, but it is much less costly and a far less intrusive design. Fingerprinting, because it is a lossy hash (compression) of thread history, is also subject to a small probability of aliasing, in which an incorrect thread history just so happens to hash to the correct thread history.

Partial Thread Replication. Some extensions of redundant multithreading have explored the ability to only partially replicate the active thread. Sundaramoorthy et al. [82] developed the Slipstream core, which provides some of the error detection of redundant multithreading but at a performance that is greater than a single thread operating alone on the core. Their key observation is that a partially redundant A-thread can run ahead of the original R-thread and act as a branch predictor and prefetcher that speeds up the execution of the R-thread compared to having the R-thread run alone. The construction of the A-thread involves removing instructions from the original thread, and this removal is performed in the compiler using heuristics that effectively guess which instructions are most helpful for generating predictions for the R-thread. Removing fewer instructions from the A-thread enables it to predict more instructions and provides better error detection (because more instructions are executed redundantly), but it also makes the A-thread take longer to execute and thus less likely to run far enough ahead of the R-stream to be helpful.

Gomaa and Vijaykumar [30] kept the A-thread intact and instead explicitly explored the tradeoff between the completeness of the R-thread, performance, and error detection coverage. They observed that the amount of redundancy can be tuned at runtime and that there are often times when redundancy can be achieved at minimal performance loss. For example, when the A-thread misses in the L2 cache, the core would otherwise be partially or mostly idle without R-thread instructions to keep it busy. They also observe that, instead of replicating each instruction in the A-thread, they can memoize (i.e., remember) the value produced by an instruction and, when that instruction is executed again, compare it to the remembered value.

The SlicK scheme of Parashar et al. [55] also provides partial replication of the A-thread. For each store instruction, if either the address or the store value predictor produces a misprediction, SlicK considers that an indication of a possible error that should be checked. In this situation, SlicK replicates the backward slice of instructions that led to this store instruction.

Redundant Threads on Multiple Cores. Redundant multithreading can be applied to system models other than SMT cores. The redundant threads can run on different cores within a multicore processor or on different cores that are on different chips. In this section, we discuss multicore processors, and we discuss multichip systems in "Redundant Multithreading on Multiple Chips" from Section 2.2.4.

The reason for using multiple cores, rather than a single SMT core, is to avoid having the threads compete for resources on the SMT core. Mukherjee et al. [51] performed a detailed simulation study of redundant multithreading, using a commercial-grade simulator of an SMT Compaq Alpha 21464 core [25]. They discovered that redundant multithreading had more of a performance impact than previously thought, and they proposed a few optimizations to help mitigate performance bottlenecks. They then proposed performing redundant multithreading on a multicore processor instead of on a single SMT core. By using separate, non-SMT cores, they avoid the core resource contention caused by having the redundant threads share the core. This design point differs from lockstepped redundant cores (Section 2.2.3) in that the redundant threads are not restricted to operating in lockstep. They show that this design point outperforms lockstepped redundant cores, by avoiding certain performance penalties inherent in lockstepping.

LaFrieda's DCC technique [36] uses redundant threads on multiple cores, but it removes the need for dedicated hardware channels for the A-thread to communicate its results to the R-thread. DCC uses the existing interconnection network to carry this traffic.

One challenge with redundant multithreading on a multicore processor is handling how the threads interact with the memory system. The threads perform loads and stores, and these loads and stores must be the same for the threads during error-free execution. There are two design options. The first option is for the threads to share the same address space. In this case, a load instruction in the A-thread may return a different value than the same load instruction in the R-thread, even during error-free execution. There are several causes of load value discrepancies, including differing observations of a cache coherence invalidation from another thread. Consider the case in which both threads load from address B. If the A-thread loads B before the R-thread loads B, it is possible for a cache coherence invalidation (requested by another thread that wishes to store a new value to B) to occur between these two loads. In this case, the R-thread's load will likely obtain a different value of B than that returned by the A-thread's load of B. There are several solutions to this problem, including having the A-thread pass the values it loads to the R-thread instead of having the R-thread perform loads. A less intrusive solution proposed by Smolens et al. [75] is to let the R-thread perform loads, detect those rare instances when interference occurs (i.e., the R-thread's load value differs from that of the A-thread), and recover to a mode in which forward progress is guaranteed.

The second option for handling how the threads interact with the memory system is for the threads to have separate memory images. This solution is conceptually simpler, and there are no problems with the threads' loads obtaining different values in error-free execution, but this solution requires software support and may waste memory space.

Redundant Multithreading on Multiple Chips. The motivation for running the redundant threads on different chips is to tolerate faults that affect a large portion of a single chip. If both threads are on a single chip that fails completely, then the error is unrecoverable. If the threads are on separate chips, the state of the thread on the chip that did not fail can be used to recover the state of the application.

The most recent Hewlett Packard NonStop machine, the NonStop Advanced Architecture (NSAA), uses redundant threads on multiple cores of multiple chips [7]. An NSAA system consists of several multiprocessors, and each thread is replicated on one core of every multiprocessor. To avoid the need for lockstepping and to reduce communication overheads between chips, the threads only compare their results when they wish to communicate with the outside world.

Similar to the case for redundant threading across cores in a multicore (*Redundant Threads on Multiple Cores*), we must handle the issue of how threads interact with the memory system. The possible solutions are the same, but the engineering tradeoffs may be different due to different costs for communication across chips.

2.2.5   Dynamic Verification of Invariants

Rather than replicate a piece of hardware or a piece of software, another approach to error detection is *dynamic verification*. At runtime, added hardware checks whether certain invariants are being satisfied. These invariants are true for all error-free executions and thus dynamically verifying them detects errors. The key to dynamic verification is identifying the invariants to check. As the invariants become more end-to-end, checking them provides better error detection (but may also incur the downsides of end-to-end error detection discussed in Section 2.1.4). Ideally, if we identify a set of invariants that completely defines correct behavior, then dynamically verifying them provides comprehensive error detection. That is, no error can occur that will not lead to a violation of at least one invariant, and thus, checking these invariants enables the detection of all possible errors. We present work in dynamic verification in an order that is based on a logical progression of invariants checked rather than in chronological order of publication.

Control Logic Checking. Detecting errors in control logic is generally more difficult than detecting errors in data because data errors can be simply detected with EDCs. In this section, we discuss dynamic verification of invariants that pertain to control.

Kim and Somani [35], in one of the first pieces of work on efficient control checking, observed that a subset of the control signals generated in the process of executing certain instructions are statically known. That is, for a given instruction, some of the control signals are always the same. To detect errors in these control signals, the authors add logic to compute a fixed-length signature of these control signals, and the core compares this signature to a prestored signature for that instruction. The prestored signature is the "golden" reference. If the runtime signature differs from the golden signature, then an error has occurred.

A related, but more sophisticated, scheme for control logic checking was developed by Reddy and Rotenberg [59]. They added a suite of microarchitectural checkers to check a set of control invariants. Similar to Kim and Somani, they added hardware to compute signatures of control signals. However, instead of computing signatures at an instruction granularity, Reddy and Rotenberg's hardware produces a signature over a trace of instructions. The core compares the runtime signature to the signature generated the last time that trace of instructions was encountered if at all. A difference indicates an error, although it is unclear which signature is the correct one. In addition to checking this invariant, their hardware checks numerous other invariants, including those pertaining to branch prediction, register renaming, and program counter updating.

The sets of invariants in this section are somewhat ad hoc in that they do not correspond to any high-level behavior of the core. They provide good error detection coverage, but they are not comprehensive. In the next several sections, we discuss sets of invariants that do correspond to high-level behaviors and that provide more comprehensive error detection.

Control Flow Checking. One high-level invariant that can be checked is that the core is faithfully executing the program's expected control flow graph (CFG). The CFG represents the sequence of instructions executed by the core, and we illustrate an example in Figure 2.7. A control flow checker [22, 42, 66, 68, 90, 92] compares the statically known CFG generated by the compiler and embedded in the program to the CFG that the core follows at runtime. If they differ, an error has been detected. A control flow checker can detect any error that manifests itself as an error in control flow. Because much of a core is involved in control flow—including the fetch, decode, and branch prediction logic—a control flow checker can detect many possible errors. To detect liveness errors in addition to safety errors, some control flow checkers also include watchdog timers that detect when no activity has occurred for a long period.

There are several challenges in implementing a control flow checker. Most notably, there are three types of instructions—data-dependent conditional branches, indirect jumps, and returns—that make it impossible for the compiler to know a priori the entire CFG of a program. The common solution to this problem is to instead check that transitions between basic blocks are correct.

Consider the example in Figure 2.8. The compiler associates a pseudounique identifier with each basic block, and it embeds in each basic block both its identifier as well as the identifiers of all of its possible successor basic blocks. Assume that the core branches from the end of A to the

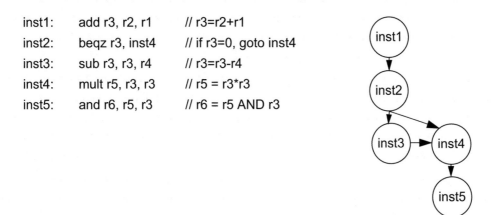

inst1:	add r3, r2, r1	// r3=r2+r1
inst2:	beqz r3, inst4	// if r3=0, goto inst4
inst3:	sub r3, r3, r4	// r3=r3-r4
inst4:	mult r5, r3, r3	// r5 = r3*r3
inst5:	and r6, r5, r3	// r6 = r5 AND r3

FIGURE 2.7: Example of CFG.

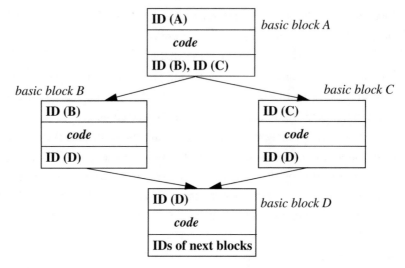

FIGURE 2.8: Control flow checking example.

beginning of B. The core then compares the identifier at B with the identifiers that were embedded at the end of A. In the error-free scenario, these identifiers are equal. An important limitation of control flow checkers is that they cannot detect if a transition is made from a basic block to the wrong successor basic block. In our example, if an error caused the core to go from A to C, the control flow checker would not detect an error because C's identifier matches the embedded identifier for C.

Another implementation challenge for control flow checkers is embedding the basic block identifiers in the program. The data flow checker can embed these identifiers into the code itself, often by inserting special NOP instructions to hold them. The drawbacks to this approach are extra instruction-cache pressure and the performance impact of having to fetch and decode these identifiers. The other option is to put the identifiers in dedicated storage. This solution has the drawback of requiring extra storage and managing its contents.

Data Flow Checking. Analogous to control flow checking, a core can also check that it is faithfully executing the data flow graph (DFG) of a program. We illustrate an example of a DFG in Figure 2.9. A data flow checker [47] embeds the DFG of each basic block in the program and the core, at runtime, computes the DFG of the basic block it is executing. If the runtime and static DFGs differ, an error is detected. A data flow checker can detect any error that manifests itself as a deviation in data flow and can thus detect errors in many core components, including the reorder buffer, reservation stations, register file, and operand bypass network. Note that a data flow checker must not only check the *shape* of the DFG but also the *values* that traverse its arcs. Fortunately, EDCs can be used to check values.

inst1:	add r3, r2, r1	// r3=r2+r1
inst2:	sub r3, r3, r4	// r3=r3-r4
inst3:	mult r5, r3, r2	// r5 = r3*r2
inst4:	and r6, r5, r3	// r6 = r5 AND r3

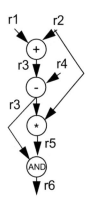

FIGURE 2.9: Example of DFG.

Data flow checking faces many of the same implementation challenges as control flow checking, including unknown branch directions and how to embed DFGs into the program. The possible solutions to these challenges are similar. One additional challenge for data flow checkers is that the size of the DFG is unbounded. To constrain the DFG size for the purposes of data flow checking, the DFG can be hashed into a fixed-length signature.

Argus. Meixner et al. [44] observed that a von Neumann core has only four tasks that must be dynamically verified: control flow, data flow, computation, and interacting with memory. They formally proved that dynamically verifying these four tasks is complete, in the absence of interrupts and I/O; that is, dynamic verification of these four tasks will detect any possible error in the core. They developed the Argus framework, which consists of checkers for each of these tasks, and they developed an initial implementation called Argus-1. Argus-1 combines existing computation checkers (like those mentioned in Section 2.2.1) with a checker for memory interactions and a checker that integrates control flow and data flow checking into one unit.

There is a synergy between control flow and data flow checking in that the DFG signatures can be used as the pseudounique basic block identifiers required for the control flow checker. To fully merge these two checkers, the compiler embeds into each basic block the DFG signatures of its possible successor basic blocks. Consider the example in Figure 2.10. If basic block A can be followed by B or C, then A contains the DFG signatures of B and C. Assume for now that the error-free scenario leads to B instead of C. When the core completes execution of B, it compares the DFG signature it computed for B to the DFG signatures that were passed to it from A. Because A passed B's DFG signature, the checker does not believe an error has occurred.

Argus-1 achieves near-complete error detection, including errors due to design bugs, because its checkers are not the same as the hardware being checked. Argus-1's error detection limitations are due to errors that occur during interrupts and I/O and errors that are undetected because its

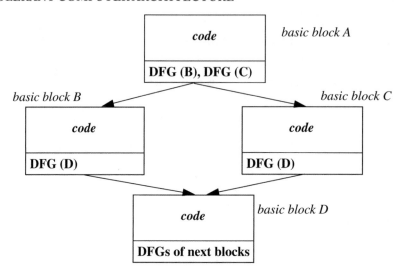

FIGURE 2.10: Integrated control flow and data flow checking example.

checkers use lossy signatures. Signatures represent a large amount of data by hashing it to a fixed-length quantity. Because of the lossy nature of hashing, there is some probability of aliasing, that is, an incorrect history happens to hash to the same value as the correct history, similar to the case for the modulo multiplier checker in "Multipliers" in Section 2.2.1. The probability of aliasing can be made arbitrarily small, but nonzero, by increasing the size of the signatures. The costs of Argus-1 are the hardware for the checkers and the power this hardware consumes. Argus-1 also introduces a slight performance penalty due to embedding the DFG signatures in the code itself.

DIVA. The first paper to introduce the term *dynamic verification* was Austin's DIVA [5]. This influential work inspired a vast amount of subsequent research in invariant checking. DIVA, like the subsequently developed Argus, seeks to dynamically verify the core. DIVA's approach, though, is entirely different from Argus. DIVA uses heterogeneous physical redundancy. It detects errors in a complex, speculative, superscalar core by checking it with a core that is architecturally identical but microarchitecturally far simpler and smaller. The checker core is a simple, in-order core with no optimizations. Because both cores have the same instruction set architecture (ISA), they produce the same results in the error-free scenario; they just produce these results in different fashions. The key to enabling the checker core to not become a throughput bottleneck is that, in the error-free scenario, the superscalar core acts as a perfect branch predictor and prefetcher for the checker core. Another throughput optimization is to use multiple checker cores in parallel. There is still a possibility of stalls due to the checkers, but these are fairly rare.

DIVA provides many benefits at low cost. The error detection coverage is excellent and it also includes errors due to design bugs in the superscalar core because the redundancy is heterogeneous.

The checker core is so simple that it can be formally verified to be bug-free, so no design bugs cause errors in it. The checker is only 6% of the area of an Alpha 21264 core [91], and the performance impact of DIVA is minimal. Comparing DIVA to Argus, DIVA achieves slightly better error detection coverage. However, DIVA is far more costly when applied to small, simple cores, instead of superscalar cores, because the checker core becomes similar in size to the core it is checking.

Watchdog Processors. Most of the invariant checkers we have discussed so far have been tightly integrated into the core. An alternative implementation is a watchdog processor, as proposed by Mahmood and McCluskey [42]. A watchdog processor is a simple coprocessor that watches the behavior of the main processor and detects violations of invariants. As illustrated in Figure 2.11, a typical watchdog shares the memory bus with the main processor. The invariants checked by the watchdog can be any of the ones discussed in this section, and the original, seminal work by Mahmood and McCluskey checked many invariants, including control flow and memory access invariants.

2.2.6 High-Level Anomaly Detection

The end-to-end argument [64], which we discussed in Section 2.1.4, motivates the idea of detecting errors by detecting when they cause higher-level behaviors that are anomalous. In this section, we present anomaly detection techniques, and we present them from the lowest-level behavioral anomalies to the highest.

Data Value Anomalies. The value of a given datum often remains constant or within a narrow range of values during the execution of a program, and an aberration from this usual behavior is likely to indicate an error. The expected range of values can be obtained either by statically profiling the program's behavior or by dynamically profiling it at runtime and inferring that this behavior is likely to continue. For example, dynamic behavior might reveal that a certain integer is always less than five. If this invariant is inferred and checked, then a subsequent assignment to this integer of the value eight would be flagged as an error. The primary challenge with such likely invariants is the possibility of false positives, that is, detecting "errors" that are not really errors but rather

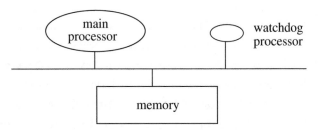

FIGURE 2.11: High-level illustration of system with watchdog processor.

violations of false invariants. Just because profiling shows that the integer is always less than five does not guarantee that some future program input could not cause it to be greater than five. Racunas et al. [58] explored several data value anomaly detectors, including those that check data value ranges, data bit invariants, and whether a data value matches one of a set of recent values. Pattabiraman et al. [57] used profiling to identify likely value invariants, and they synthesize hardware that can efficiently detect violations of these value invariants at runtime.

Microarchitectural Behavior Anomalies. Data value anomalies represent one possible type of anomaly to detect, and they are still fairly low-level anomalies. At a higher level, one can detect microarchitectural behaviors that are anomalous. Wang and Patel's ReStore [89] architecture detects transient errors by detecting microarchitectural behaviors that, although possible in an error-free execution, are rare enough to be suspicious. These behaviors include exceptions, page faults, and branch mispredictions that occur despite the branch confidence predictor having high confidence in the predictions. All of these behaviors may occur in error-free execution, but they are relatively infrequent. If ReStore observes any of these behaviors, it recovers to a pre-error checkpoint and replays execution. If the anomalous behavior does not recur during replay, then it was most likely due to a transient error. If it does recur, then it was either a legal but rare behavior or it is due to a permanent fault.

Software Behavior Anomalies. One consequence of the end-to-end argument is that detecting hardware errors when they affect software behavior is, if possible, preferable to detecting these errors at the hardware level. Intuitively, an error only matters if it affects software behavior, and detecting hardware errors that do not impact the software is not necessary. Computer users do not notice if a transistor fails or a bit of SRAM is flipped by a cosmic ray; they notice when their programs crash.

The SWAT system of Li et al. [40] exploits this observation to achieve low-cost error detection for cores. Certain software behaviors are atypical of error-free operation and are likely to result from either a hardware error or a software bug; SWAT focuses on the hardware errors. These suspicious software behaviors include fatal exceptions, program crashes, an unusually high amount of operating system activity, and hangs. All of these behaviors are easily detectable with minimal extra hardware or software.

SWAT adheres to the end-to-end argument and achieves its benefits: low additional hardware and software costs, little performance overhead, no false positives (detecting errors that do not affect the software), and the potential for comprehensive error detection. SWAT is not comprehensive, though, because some hardware errors do not manifest themselves in software behaviors that SWAT detects. These errors cause silent data corruptions that violate safety. One example of such an error is an error that corrupts a floating point unit's computation. In many cases, such an error will not cause the software to obviously misbehave. In theory, one could extend SWAT with more software checks to detect these errors, but one must be careful that such an approach does not devolve into self-checking code [10], with its vastly greater performance overhead than SWAT.

Software-level error detection has the expected drawbacks of end-to-end error detection that were discussed in Section 2.1.4. First, there is no bound on how long it may take for a hardware error to manifest itself at the software level. The latency between the occurrence of the hardware error and its detection is thus unbounded, although in practice it is usually reasonably short. Nevertheless, SWAT's error detection latency is significantly longer than that of a hardware-level error detection scheme. Second, when SWAT detects an error, it can provide little or no diagnostic information. The group that developed SWAT added diagnostic capability to it in subsequent work [39] that we discuss in Chapter 4.

2.2.7 Using Software to Detect Hardware Errors

All of the previous error detection schemes we have presented have primarily used hardware to detect errors in the core. The control flow and data flow checkers and Argus used some compiler help to embed signatures into the program, but still most of error detection was performed in hardware. SWAT used mostly simple hardware checks with a little additional software. We now change course a bit and explore some techniques for using software to detect errors in the core.

One approach to software-implemented detection of hardware errors is to create programs that have redundant instructions in them. One of the first approaches to this was the error detection by duplicated instructions (EDDI) of Oh et al. [54]. The key idea was to insert redundant instructions and also insert instructions that compare the results produced by the original instructions and the redundant instructions. We illustrate a simple example of this approach in Figure 2.12. The SWIFT scheme of Reis et al. [61] improved upon the EDDI idea by combining it with control flow checking (Control Flow Checking from Section 2.2.5) and optimizing the performance by reducing the number of comparison instructions.

The primary appeal of software redundancy is that it has no hardware costs and requires no hardware design modifications. It also provides good coverage of possible errors, although it has some small coverage holes that are fundamental to all-software schemes. For example, consider a store instruction. If the store is replicated and the results are compared by another instruction, the core can be sure that the store instruction has the correct address and data value to be stored. However, there is no way to check whether either the address or data are corrupted between when the comparison instruction completes and when the store's effect actually takes place on the cache. Another problematic error model is a multiple-error scenario in which one error causes one of the two redundant instructions to produce the wrong result and another error causes the comparison instruction to either not occur or mistakenly believe that the redundant instructions produced the same result.

Original Code		Code with EDDI-like Redundancy	
add r1, r2, r3	// r1 = r2 + r3	add r1, r2, r3	// r1 = r2 + r3
xor r4, r1, r5	// r4 = r1 XOR r5	add r11, r12, r13	// r11 = r12 + r13
store r4, 0($r6)	// Mem[$r6] = r4	xor r4, r1, r5	// r4 = r1 XOR r5
		xor r14, r11, r15	// r14 = r11 XOR r15
		bne r4, r14, error	// if r4 !=r14, goto error
		store r4, 0($r6)	// Mem[$r6] = r4

FIGURE 2.12: EDDI-like software-implemented error detection. The redundant code is compared before the store instruction.

The costs of software redundancy are significant. The dynamic energy overhead is more than 100%, and the performance penalty is also substantial. The performance penalty depends on the core model and the software workload on that core—a wide superscalar core executing a program with little instruction-level parallelism will have enough otherwise unused resources to hide much of the latency of executing the redundant instructions. However, a narrower core or a more demanding software workload can lead to performance penalties on the order of 100%; in the extreme case of a 1-wide core that would be totally used by the nonredundant software, adding redundant instructions would more than double the runtime.

2.2.8   Error Detection Tailored to Specific Fault Models

Many of the error detection schemes we have discussed in this chapter have had fairly general error models. They all target transient errors, and many also detect errors due to permanent faults and perhaps even errors due to design bugs. In this section, we discuss error detection techniques that are specifically tailored for errors due to permanent faults and design bugs but do not target transient errors.

Errors Due to Permanent Faults. Recent trends that predict an increase in permanent wear-out faults [80] have motivated schemes to detect errors due to permanent faults and diagnose their locations.

Blome et al. [8] developed wear-out detectors that can be placed at strategic locations within a core. The key observation is that wear-out of a component often manifests itself as a progressive increase in that component's latency. They add a small amount of hardware to statistically assess increases in delay and thus detect the onset of wear-out. A component with progressively increasing delay is diagnosed as wearing out and likely to soon suffer a permanent fault.

Instead of monitoring a set of components for increasing delay, the BulletProof approach of Shyam et al. [72] performs periodic built-in self-test (BIST) of every component in the core. During each "computation epoch," which is the time between taken checkpoints, the core uses spare cycles to perform BIST (e.g., testing the adder when the adder would otherwise be idle). If BIST identifies a permanent fault, then the core recovers to a prior checkpoint. If BIST does not identify any permanent faults, then the computation epoch was executed on fault-free hardware and a checkpoint can be taken that incorporates the state produced during that epoch. Constantinides et al. [21] showed how to increase the flexibility and reduce the hardware cost of the BulletProof approach by implementing the BIST partially in software. Their scheme adds instructions to the ISA that can access and modify the scan chain used for BIST; using these instructions, test programs can be written that have the same capability as all-hardware BIST.

Smolens et al. [76] developed a scheme, called FIRST, that cleverly integrates ideas from both Blome et al. and BulletProof. Periodically, but far less frequently than BulletProof, FIRST performs BIST. Unlike BulletProof, which detects permanent faults, the goal of this BIST is to uncover wear-out before it leads to permanent faults. FIRST performs the BIST at various clock frequencies to observe at which frequency the core no longer meets its timing requirements. If this frequency progressively decreases, it is likely a sign of wear-out and an imminent hard fault.

Errors Due to Design Bugs. Errors due to design bugs are particularly problematic because a design bug affects every shipped core. The infamous floating point division bug in the Intel Pentium [11] led to an extremely expensive recall of all of the shipped chips. Unfortunately, design bugs will continue to plague shipped cores because completely verifying the design of a complicated core is well beyond the current state-of-the-art in verification technology. Ideally, we would like a core to be able to detect errors due to design bugs and, if possible, recover gracefully from these errors.

Wagner et al. [87], Narayanasamy et al. [52], and Sarangi et al. [65] take similar approaches to detecting errors due to design bugs. They assume that the bugs have already been discovered—either by the manufacturer or by consumers who report the problem to the manufacturer—and that the manufacturer has communicated a list of these bugs to the core. They observe that matching these bugs to dynamic core behaviors requires the core to monitor only a relatively small subset of its internal signals. Their schemes monitor these signals and continuously compare them, or their signature, to known values that indicate that a design bug has manifested itself. If a match occurs, the core has detected an error and can try to recover from it, perhaps by using a BIOS patch or some other workaround.

Constantinides et al. [20] have the same goal of detecting errors due to design bugs, but they make two important contributions. First, they use an RTL-level analysis, rather than the previously used microarchitectural analysis, to show that far more signals than previously reported must be monitored to detect errors due to design bugs. Second, they present an efficient scheme for monitoring every control signal, rather than just a subset. They observe that they must monitor only flip-flops, and they use the preexisting scan flip-flop that corresponds to each operational flip-flop to hold a bit that indicates whether the operational flip-flop must be monitored. They augment each operational flip-flop with a flip-flop that holds the data value to be matched for that operational flip-flop.

2.3 CACHES AND MEMORY

Error detection for processor cores has historically existed only in high-end computers, although trends suggest that more error detection is likely to be necessary in future commodity cores. However, another part of the computer, the storage, has commonly had error detection even in inexpensive commodity computers. There are three reasons why caches and memory have historically featured error detection despite a relative lack of error detection for the cores.

First, the DRAM that comprises main memory has long been known to be susceptible to transient errors [96], and the SRAM that comprises caches has been more recently discovered to be susceptible. Historically, DRAM and SRAM have been orders of magnitude more susceptible than logic to transient errors, although this relationship is quickly changing [71].

Second, caches and memory represent a large fraction of a processor. The size of memory has grown rapidly, to the point where even a laptop may have a few gigabytes. Also, as Moore's Law has provided architects with more and more transistors per chip, one trend has been to increase cache sizes. Given a constant rate of errors per bit, which is unrealistically optimistic, having more bits in a cache or memory presents more opportunities for errors.

Third, and perhaps most importantly, there is a simple and well-understood solution for detecting (and correcting) errors in storage: error detecting (and correcting) codes. EDC provides an easily understood error detection capability that can be adjusted to the anticipated error model, and it has thus been incorporated into most commercial computer systems. In most computers, the levels of the memory hierarchy below the L1 caches, including the L2 cache and memory, are protected with ECC. The L1 cache is either protected with EDC (as in the Pentium 4 [31], Ultra-SPARC IV [81], and Power4 [13]) or with ECC (as in the AMD K8 [1] and Alpha 21264 [33]).

2.3.1 Error Code Implementation

The choice of error codes represents an engineering tradeoff. Using EDC on an L1 cache, instead of ECC, leads to a smaller and faster L1 cache. However, with only EDC on the L1, the L1 must be write-through so that the L2 has a valid copy of the data if the L1 detects an error. The write-through L1 consumes more L2 bandwidth and power compared to a write-back L1.

Some recent research attempts to achieve the best of both worlds. The punctured ECC recovery cache (PERC) [63] uses a special type of ECC, called a punctured code, that enables the redundant bits necessary for error detection to be kept separately from the additional redundant bits necessary for error correction. By keeping the bits required for error detection in the L1 and the additional bits for correction in a separate structure, the L1 remains small and fast in the common, error-free case.

Other error coding schemes for caches and memories are tailored to particular error models. For example, spatially correlated errors are difficult for many error coding schemes because a typical code is designed to tolerate one or maybe two errors per word or block (where the error code is applied at the granularity of a word or block). One option to tolerate spatially correlate errors is to interleave bits from different words or blocks such that an error in several spatially close bits does not affect more than one bit (or a small number of bits) per word or block. For main memory, which often consists of multiple DRAM chips, this interleaving can be done at many levels, including across banks and chips. Interleaving across chips protects the memory from a chipkill failure of a single DRAM chip.

For caches, a more efficient and scalable approach to error coding for spatially correlated errors is a two-dimensional coding scheme proposed by Kim et al. [34]. Their scheme applies EDC on each row of the cache and thus maintains fast error-free accesses, similar to the PERC. The twist is that they compute an additional error code over the columns of the cache. If an error is detected in a row, the column's error code can be accessed to help correct it. With this organization of the redundant bits, they can efficiently tolerate large spatial errors without adding to the latency of error-free accesses.

2.3.2   Beyond EDCs

Because of the importance of detecting errors in caches, there has recently been work that has gone beyond simple EDC and ECC.

One previously known idea that has reemerged in this context is scrubbing [50]. Scrubbing a memory structure involves periodically reading each of its entries and detecting (and/or correcting) any errors found in these accesses. The purpose of scrubbing is to remove latent errors before they accumulate beyond the capabilities of the EDC. Consider a cache that uses parity for error detection. Assume that errors are fairly rare and only occur in one bit at a time. In this situation, parity appears sufficient for detecting errors. However, consider a datum that has not been accessed for months. Multiple errors might have occurred in that time frame, thus violating our single-error assumption and making parity insufficient. Cache scrubbing bounds the maximum time between accessing each datum and thus avoids these situations. In industry, AMD's recent processors provide examples of processors that use scrubbing for caches and memory [4].

A more radical approach to cache error detection is In-Cache Replication (ICR) [95]. The idea behind ICR is to use otherwise unoccupied, invalid cache frames to hold replicas of data that are held in other parts of the cache. Comparing a replica to the original datum enables error detection. More sophisticated uses of ICR use the replica to aid in error correction as well. The ICR work was followed by the replication cache idea [94] that enabled replicas to reside in a small, dedicated structure, instead of occupying valuable cache frames.

2.3.3   Detecting Errors in Content Addressable Memories

Most storage structures are randomly accessible by address. For these structures, an address is applied to the structure, and that address within the structure is read or written. However, another important class of storage that we must consider is the content addressable memory (CAM). A CAM is a collection of names that can be matched by an input name. A CAM is read by providing it with a name and then the CAM responds with the locations of the entries that match that input name. CAMs are useful structures, and they are commonly used in caches, among other purposes. A common cache organization, shown in Figure 2.13, uses a CAM to hold the tags. Each CAM entry corresponds to a RAM entry that holds the data corresponding to that tag. If an address matches a name in the CAM, that CAM entry outputs a one and accesses the corresponding data from the matching RAM entry. If the address does not match any name in the CAM, the CAM responds that the address missed in the cache.

A problematic error scenario for CAMs is an error in an entry's name field. Assume there is an entry that should be <B, 3>. In the error-free case, a read of B returns the value 3. However, an error in the name field, which say changes the entry to <C, 3>, may lead to two possible problems. Assume that, in the error-free case, there is no entry with the name C. The first problem is that accessing the CAM with name B will not return the value 3. The second problem is that accessing the CAM with name C will erroneously return the value 3, when it should have returned a miss.

If a CAM is being used in a cache, these two problem scenarios are equivalent to false misses and false hits, respectively, both of which can violate safety. Assume the cache is a write-back L1 data cache and that the data value of address B in the cache is more recent than the value of address B in the L2 cache. A false miss will cause the core to access the L2 cache and return stale data for address B. The false-miss problem does not violate safety for write-though caches because the L2 will have the current data value of B. A false hit will provide the core with erroneous data for an access to address C.

At first glance, it might appear that simply protecting the CAM entries with parity or some other EDC would be sufficient. However, consider our example again and assume that the EDC protected version of B is EDC(B). The CAM entry should hold EDC(B), but it instead holds some erroneous value that we will assume happens to be EDC(C). If we access the CAM with the input

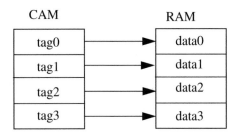

FIGURE 2.13: Cache organization using CAM.

EDC(B), we will still have a false miss because EDC(B) does not match C. If we access the CAM with the input EDC(C), we will still have a false hit. The reason the errors are undetected is that most CAMs just perform a match but, for efficiency reasons, do not explicitly inspect the entries that are being matched.

The key to using EDC to detect CAM errors is to modify the comparison logic to explicitly inspect the CAM entries. The scheme of Lo [41] adds EDC to each entry in the CAM and then modifies the comparison logic to detect both false misses and false hits. Assume for purposes of this explanation that the EDC is parity and that the error model is single-bit flips. If the CAM entry is identical to the input name, then it is a true hit; there is no way for an input name to match a CAM entry that has a single-bit error. False hits are impossible. If the CAM entry differs from the input name in more than one bit position, this is a true miss because all true misses will differ in at least two bit positions. If the CAM entry differs from the input name in exactly one bit position, then this is a false miss. This approach can be extended to EDCs other than parity.

2.3.4 Detecting Errors in Addressing

One subtle error model for memory structures is the situation in which the memory has faulty addressing. Consider the case where a core accesses a memory with address B, and the memory erroneously provides it with the correct data value *at address C*. Even with EDC, this error will go undetected because the data value at address C is error-free. The problem is not the value at address C; rather, the problem is that the core wanted the value at address B. EDC only protects the data values.

Meixner and Sorin [44] developed a way to detect this error as part of Argus's memory checker. The key is to embed the address with the datum. Conceptually, one can imagine keeping a complete copy of the address with each datum. This solution would work, but it would require a huge amount of extra storage for the addresses. Instead, they embed the address in the EDC of the data, as shown in an example in Figure 2.14. When storing value D to address A, the core writes D along with [EDC(D XOR A)] in that location. When the core reads address A and obtains

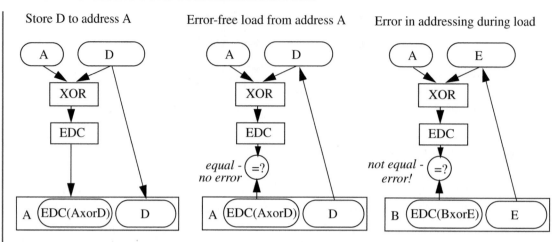

FIGURE 2.14: Detecting errors in addressing.

the value D, it compares the expected EDC, which is EDC(D XOR A) with the EDC that was returned along with D. These two EDC values will be equal if there is no error. However, consider the case where the core wishes to read A, but an error in memory addressing causes the memory to return the contents at address B, which are the values E and EDC(E XOR B). Because EDC(E XOR B) does not equal EDC(E XOR A), except in extremely rare aliasing situations, an error in addressing is detected.

2.4   MULTIPROCESSOR MEMORY SYSTEMS

Multiprocessors, including multicore processors, have components other than the cores and the memory structures themselves. A multiprocessor's memory system also includes the interconnection network that enables the cores to communicate and the cache coherence hardware. These memory systems are complicated distributed systems, and detecting errors in them is challenging. One particular challenge is that detecting errors in each individual component may not be sufficient because we must also detect errors in the interactions between the components. Furthermore, some errors may be extremely difficult to detect with a collection of strictly localized, per-component checkers because the error only manifests itself as a violation of a global invariant.

As an example of a difficult-to-detect error, consider a multicore processor in which the cores are connected with a logical bus that is implemented as a tree (like the Sun UltraEnterprise E10000 [16]), as shown in Figure 2.15. The cores use a snooping cache coherence protocol that relies on cache coherence requests being totally ordered by the logical bus. Core 1 and core 2 broadcast cache coherence requests by unicasting their requests to the root of the tree. The winner, core 1, has its request broadcast down the tree, followed by core 2's request. In the error-free case, all cores observe core 1's request before core 2's request, and the coherence protocol works correctly. Now assume that

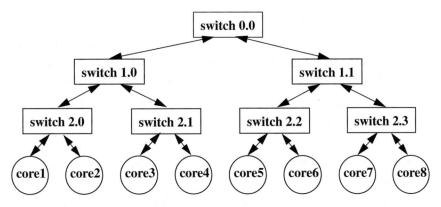

FIGURE 2.15: Example system: multicore processor with logical bus implemented as tree.

an error in switch 2.2 reorders the requests as observed by core 5. This error will lead to a violation of coherence, yet it is very difficult to detect. The requests arrive uncorrupted at core 5, so their EDC checks do not reveal an error. A timeout mechanism would not work because the requests reach every core and thus get responses. One could argue that we should just add dedicated hardware to check for this error scenario, but then we must worry if there are other scenarios like this one that we have not considered. Or one could argue that we should just replicate the switches, but this approach is costly.

Challenging error models like this one have motivated the use of dynamic verification of end-to-end invariants rather than attempting to create dedicated hardware checkers for every possible component and error model. These schemes are the focus of the rest of this chapter, and they are an emerging area of research, as compared to the long history of error detection schemes for cores.

2.4.1   Dynamic Verification of Cache Coherence

Cache coherence is a global invariant that lends itself to dynamic verification. Coherence is a required property, and an error-free memory system maintains it at all times. Dynamic verification of cache coherence can detect any error that manifests itself as a violation of coherence. We present work in this area chronologically, to show the progression of ideas.

Cantin et al. [15] first identified dynamic verification of cache coherence as an attractive way to detect errors in memory systems. Their implementation was inspired by the DIVA scheme [5] (*DIVA* from Section 2.2.5) and, analogous to DIVA, it checks a complicated, high-performance coherence protocol with a simpler protocol.[1] This scheme is limited to snooping protocols, and it requires replication of the cache line state information and an additional snooping bus. The scheme achieves good error detection coverage but at steep hardware and performance costs.

---

[1] DIVA checks a complicated, high-performance core with a simpler core.

Sorin et al. [79] developed a less costly but less complete scheme for detecting errors in snooping cache coherence. They develop hardware to check two invariants that are necessary but not sufficient for achieving coherence. The first invariant is that all cores see the same total order of coherence requests. The second invariant is that all coherence upgrades have corresponding downgrades elsewhere in the system. The invariant checking hardware is cheap and the scheme has negligible performance impact, but it is limited to snooping coherence protocols and it cannot detect all errors in coherence.

Meixner and Sorin [48] developed a scheme called Token Coherence Signature Checking (TCSC) that overcomes the limitations of the first two schemes we discussed. The key idea of TCSC is to have each cache controller and memory controller compute a signature of the history of coherence events it has performed. Periodically, the signatures of every controller are aggregated at a single small checker that can determine, by examining the signatures, whether an error has occurred. By carefully choosing the signature computation functions, the hardware costs and additional interconnection network traffic are kept low. TCSC applies to any type of coherence protocol, including directory and token coherence [43]. TCSC is complete; it detects any error that affects coherence. TCSC adds little hardware and has only a small impact on performance.

Fernandez-Pascual et al. [27, 28] developed a somewhat different approach to detecting errors in snooping and directory coherence protocols. Instead of dynamically verifying coherence, they add a set of timeout mechanisms to the coherence protocol. For example, when a core initiates a coherence request, it sets a timer that, if it expires before the request is satisfied, indicates an error. By carefully choosing the actions for which to set timers, their schemes achieve excellent error detection coverage at low hardware cost. Furthermore, they augment the coherence protocol with the ability to recover itself after a timer detects an error.

The CoSMa scheme of DeOrio et al. [23] is somewhat similar in approach to TCSC, but its goals are different. It is designed for post-silicon validation purposes rather than for in-field error detection. Because it will not be used in the common case, it must use little additional hardware and it must be possible to disable it in the field. CoSMa does not need to be as fast as TCSC because it is not meant to be used in the field. CoSMa works by logging coherence events and periodically stopping the processor to analyze the logs for indications of errors. If errors are detected, they may indicate underlying design bugs that the manufacturer is trying to uncover during post-silicon validation and before shipping the product.

2.4.2 Dynamic Verification of Memory Consistency

As we have mentioned before, the key to dynamic verification is identifying the invariants to check. A more complete set of invariants enables better error detection coverage. For a memory system, the most complete invariant is the *memory consistency model* [2]. The memory consistency model formally defines the correct end-to-end behavior of the memory system; a system obeying its consistency model is behaving correctly. Thus, dynamic verification of memory consistency is sufficient for detecting any error in the memory system. As with dynamic verification of cache coherence, we present the research in this area in chronological order.

Cain and Lipasti [14] first identified dynamic verification of consistency as an appealing technique for detecting errors in the memory system. They developed an algorithm that uses vector clocks to track the orderings of reads and writes. By checking this ordering, the algorithm can determine whether the memory system is obeying its consistency model. Their algorithm is elegant, but they did not present a hardware implementation.

Meixner and Sorin [45] developed a scheme for dynamic verification of sequential consistency (DVSC). Sequential consistency (SC) [37] is a restrictive memory consistency model, in that it permits few reorderings of reads and writes. Instead of directly checking SC, DVSC checks several sub-invariants that are provably equivalent to SC. This indirect approach enables an efficient implementation. Meixner and Sorin [46] followed DVSC with dynamic verification of memory consistency (DVMC), in general. DVMC applies to a wide range of consistency models, including all commercially implemented consistency models. Like DVSC, DVMC takes an indirect approach in which the memory consistency invariant is divided into sub-invariants that are checked. DVMC's sub-invariants are, however, quite different. DVMC's three sub-invariants are the following: the core behaves logically in-order, the allowable reorderings are enforced, and the caches are coherent. Checking the first two invariants is simple and requires little hardware; checking coherence can be done with any of the schemes discussed in Section 2.4.1.

Chen et al. [17] developed an implementation of DVMC that directly checks the memory consistency invariant. Their scheme records all of the orderings observed between reads and writes, not unlike Cain and Lipasti [14], and then checks that this graph contains no illegal cycles that indicate a consistency violation. The key to the implementation's efficiency is that they optimize this graph, by pruning unnecessary information, to keep it small and feasible to check at runtime. By directly checking the consistency invariant, instead of the sub-invariants checked by Meixner and Sorin's [46] approach, their scheme is applicable to an even wider range of possible memory consistency models. Chen et al. [18] followed up this work with a dynamic verification scheme that applies to memory systems that provide transactional memory.

DeOrio et al. [24] developed Dacota to dynamically verify memory ordering invariants that are necessary for memory consistency. Dacota's approach is similar to that of Chen et al. [17] in that it records read and write orderings and searches for illegal cycles in this graph of orderings. Unlike other DVMC implementations, Dacota's goal is not to detect runtime errors; rather, the goal is to use Dacota as a post-silicon validation tool. After the first silicon is produced, Dacota would detect memory ordering violations and thus uncover design bugs. Because the goal is post-silicon validation, Dacota's implementation is optimized for area. Dacota's performance impact is less important because it is disabled after the chip is shipped.

2.4.3   Interconnection Networks

There are numerous schemes for detecting errors in interconnection networks, and these schemes are generally quite similar to the approaches for detecting errors in more general networks. The two most common error detecting schemes are EDC and timeouts. Putting EDC on packets is an effective solution for detecting errors in links or switches that lead to corrupted packets. Timeouts are effective at detecting lost messages.

2.5   CONCLUSIONS

Error detection is an active and exciting field. Although many excellent techniques exist, error detection is by no means a solved problem. In particular, there are at least three interesting open problems:

- Efficient error detection for floating point units (FPUs): We are unaware of any reasonably efficient—in terms of hardware and performance overheads—schemes for detecting errors in FPUs. Duplication is currently the only viable approach for comprehensively detecting errors. Some arithmetic coding schemes can be used, but their costs are quite high.
- Error detection for multiple-error scenarios: If the forecasts of greatly increased fault rates come to pass, then error detection schemes that target single-error scenarios may be insufficient. Most of the current schemes assume a single-error model, which is reasonable today, but may not be appropriate in the future. Some existing schemes may do well at detecting multiple-error scenarios, but we are unaware of results that demonstrate this capability.
- Error detection for other processor models: It is likely that error detection schemes for other processor models, such as graphics processing units (GPUs) and network processing units, will have different requirements and engineering constraints. Dynamic verification schemes would likely require different sets of invariants. It is also unclear how much error detection is required for these models—for example, errors in GPUs that cause erroneous individual pixels are not worth detecting.

2.6   REFERENCES

[1]   Advanced Micro Devices. AMD Eighth-Generation Processor Architecture. Advanced Micro Devices Whitepaper, Oct. 2001.

[2]   S. V. Adve and K. Gharachorloo. Shared Memory Consistency Models: A Tutorial. *IEEE Computer*, 29(12), pp. 66–76, Dec. 1996. doi:10.1109/2.546611

[3]   N. Aggarwal, P. Ranganathan, N. P. Jouppi, and J. E. Smith. Configurable Isolation: Building High Availability Systems with Commodity Multi-Core Processors. In *Proceedings of the 34th Annual International Symposium on Computer Architecture*, pp. 470–481, June 2007.

[4]   AMD. BIOS and Kernel Developer's Guide for AMD Athlon 64 and AMD Opteron Processors. Publication 26094, Revision 3.30, Feb. 2006.

[5]   T. M. Austin. DIVA: A Reliable Substrate for Deep Submicron Microarchitecture Design. In *Proceedings of the 32nd Annual IEEE/ACM International Symposium on Microarchitecture*, pp. 196–207, Nov. 1999. doi:10.1109/MICRO.1999.809458

[6]   A. Avizienis and J. P. J. Kelly. Fault Tolerance by Design Diversity: Concepts and Experiments. *IEEE Computer*, 17, pp. 67–80, Aug. 1984.

[7]   D. Bernick et al. NonStop Advanced Architecture. In *Proceedings of the International Conference on Dependable Systems and Networks*, June 2005. doi:10.1109/DSN.2005.70

[8]   J. Blome, S. Feng, S. Gupta, and S. Mahlke. Self-Calibrating Online Wearout Detection. In *Proceedings of the 40th Annual IEEE/ACM International Symposium on Microarchitecture*, Dec. 2007.

[9]   J. A. Blome et al. Cost-Efficient Soft Error Protection for Embedded Microprocessors. In *Proceedings of the International Conference on Compilers, Architecture, and Synthesis for Embedded Systems*, Oct. 2006. doi:10.1145/1176760.1176811

[10] M. Blum and S. Kannan. Designing Programs that Check Their Work. In *ACM Symposium on Theory of Computing*, pp. 86–97, May 1989. doi:10.1145/73007.73015

[11] M. Blum and H. Wasserman. Reflections on the Pentium Bug. *IEEE Transactions on Computers*, 45(4), pp. 385–393, Apr. 1996. doi:10.1109/12.494097

[12] D. Boggs et al. The Microarchitecture of the Intel Pentium 4 Processor on 90nm Technology. *Intel Technology Journal*, 8(1), Feb. 2004.

[13] D. C. Bossen, J. M. Tendler, and K. Reick. Power4 System Design for High Reliability. *IEEE Micro*, 22(2), pp. 16–24, Mar./Apr. 2002.

[14] H. W. Cain and M. H. Lipasti. Verifying Sequential Consistency Using Vector Clocks. In *Revue in Conjunction with Symposium on Parallel Algorithms and Architectures*, Aug. 2002. doi:10.1145/564870.564897

[15] J. F. Cantin, M. H. Lipasti, and J. E. Smith. Dynamic Verification of Cache Coherence Protocols. In *Workshop on Memory Performance Issues*, June 2001.

[16] A. Charlesworth. Starfire: Extending the SMP Envelope. *IEEE Micro*, 18(1), pp. 39–49, Jan./Feb. 1998. doi:10.1109/40.653032

[17] K. Chen, S. Malik, and P. Patra. Runtime Validation of Memory Ordering Using Constraint Graph Checking. In *Proceedings of the Thirteenth International Symposium on High-Performance Computer Architecture*, Feb. 2008.

[18] K. Chen, S. Malik, and P. Patra. Runtime Validation of Transactional Memory Systems. In *Proceedings of the International Symposium on Quality Electronic Design*, Mar. 2008.

[19]   W. J. Clarke et al. IBM System z10 Design for RAS. *IBM Journal of Research and Development*, 53(1), pp. 11:1–11:11, 2009.

[20]   K. Constantinides, O. Mutlu, and T. Austin. Online Design Bug Detection: RTL Analysis, Flexible Mechanisms, and Evaluation. In *Proceedings of the 41st Annual IEEE/ACM International Symposium on Microarchitecture*, Nov. 2008.

[21]   K. Constantinides, O. Mutlu, T. Austin, and V. Bertacco. Software-Based Online Detection of Hardware Defects: Mechanisms, Architectural Support, and Evaluation. In *Proceedings of the 40th Annual IEEE/ACM International Symposium on Microarchitecture*, pp. 97–108, Dec. 2007.

[22]   X. Delord and G. Saucier. Formalizing Signature Analysis for Control Flow Checking of Pipelined RISC Microprocessors. In *Proceedings of International Test Conference*, pp. 936–945, 1991. doi:10.1109/TEST.1991.519759

[23]   A. DeOrio, A. Bauserman, and V. Bertacco. Post-Silicon Verification for Cache Coherence. In *Proceedings of the IEEE International Conference on Computer Design*, Oct. 2008.

[24]   A. DeOrio, I. Wagner, and V. Bertacco. DACOTA: Post-Silicon Validation of the Memory Subsystem in Multi-Core Designs. In *Proceedings of the Fourteenth International Symposium on High-Performance Computer Architecture*, Feb. 2009.

[25]   K. Diefendorff. Compaq Chooses SMT for Alpha. *Microprocessor Report*, 13(16), pp. 6–11, Dec. 1999.

[26]   E. Elnozahy and W. Zwaenepoel. Manetho: Transparent Rollback-Recovery with Low Overhead, Limited Rollback, and Fast Output Commit. *IEEE Transactions on Computers*, 41(5), pp. 526–531, May 1992. doi:10.1109/12.142678

[27]   R. Fernandez-Pascual, J. M. Garcia, M. Acacio, and J. Duato. A Low Overhead Fault Tolerant Coherence Protocol for CMP Architectures. In *Proceedings of the Twelfth International Symposium on High-Performance Computer Architecture*, Feb. 2007.

[28]   R. Fernandez-Pascual, J. M. Garcia, M. Acacio, and J. Duato. A Fault-Tolerant Directory-Based Cache Coherence Protocol for Shared-Memory Architectures. In *Proceedings of the International Conference on Dependable Systems and Networks*, June 2008.

[29]   M. A. Gomaa, C. Scarborough, T. N. Vijaykumar, and I. Pomeranz. Transient-Fault Recovery for Chip Multiprocessors. In *Proceedings of the 30th Annual International Symposium on Computer Architecture*, pp. 98–109, June 2003. doi:10.1145/859630.859631, doi:10.1145/859618.859631

[30]   M. A. Gomaa and T. N. Vijaykumar. Opportunistic Transient-Fault Detection. In *Proceedings of the 32nd Annual International Symposium on Computer Architecture*, pp. 172–183, June 2005. doi:10.1109/ISCA.2005.38

[31]   Intel. *Intel Pentium 4 Processor on 90 nm Process Datasheet*. Intel Corporation, Apr. 2004.

[32]   D. Jewett. Integrity S2: A Fault-Tolerant UNIX Platform. In *Proceedings of the 21st International Symposium on Fault-Tolerant Computing Systems*, pp. 512–519, June 1991. doi:10.1109/FTCS.1991.146709

[33] R. E. Kessler. The Alpha 21264 Microprocessor. *IEEE Micro*, 19(2), pp. 24–36, Mar./Apr. 1999. doi:10.1109/40.755465

[34] J. Kim, N. Hardavellas, K. Mai, B. Falsafi, and J. C. Hoe. Multi-Bit Error Tolerant Caches Using Two-Dimensional Error Coding. In *Proceedings of the 40th Annual IEEE/ACM International Symposium on Microarchitecture*, Dec. 2007.

[35] S. Kim and A. K. Somani. On-Line Integrity Monitoring of Microprocessor Control Logic. In *Proceedings of the International Conference on Computer Design*, pp. 314–319, Sept. 2001.

[36] C. LaFrieda, E. Ipek, J. F. Martinez, and R. Manohar. Utilizing Dynamically Coupled Cores to Form a Resilient Chip Multiprocessor. In *Proceedings of the International Conference on Dependable Systems and Networks*, June 2007.

[37] L. Lamport. How to Make a Multiprocessor Computer that Correctly Executes Multiprocess Programs. *IEEE Transactions on Computers*, C-28(9), pp. 690–691, Sept. 1979.

[38] G. G. Langdon and C. K. Tang. Concurrent Error Detection for Group Look-Ahead Binary Adders. *IBM Journal of Research and Development*, 14(5), pp. 563–573, Sept. 1970.

[39] M.-L. Li, P. Ramachandran, S. K. Sahoo, S. Adve, V. Adve, and Y. Zhou. Trace-Based Diagnosis of Permanent Hardware Faults. In *Proceedings of the International Conference on Dependable Systems and Networks*, June 2008.

[40] M.-L. Li, P. Ramachandran, S. K. Sahoo, S. Adve, V. Adve, and Y. Zhou. Understanding the Propagation of Hard Errors to Software and Implications for Resilient System Design. In *Proceedings of the Thirteenth International Conference on Architectural Support for Programming Languages and Operating Systems*, Mar. 2008. doi:10.1145/1346281.1346315

[41] J.-C. Lo. Fault-Tolerant Content Addressable Memory. In *Proceedings of the IEEE International Conference on Computer Design*, pp. 193–196, Oct. 1993. doi:10.1109/ICCD.1993.393382

[42] A. Mahmood and E. McCluskey. Concurrent Error Detection Using Watchdog Processors—A Survey. *IEEE Transactions on Computers*, 37(2), pp. 160–174, Feb. 1988. doi:10.1109/12.2145

[43] M. M. K. Martin, M. D. Hill, and D. A. Wood. Token Coherence: Decoupling Performance and Correctness. In *Proceedings of the 30th Annual International Symposium on Computer Architecture*, June 2003. doi:10.1109/ISCA.2003.1206999

[44] A. Meixner, M. E. Bauer, and D. J. Sorin. Argus: Low-Cost, Comprehensive Error Detection in Simple Cores. In *Proceedings of the 40th Annual IEEE/ACM International Symposium on Microarchitecture*, pp. 210–222, Dec. 2007.

[45] A. Meixner and D. J. Sorin. Dynamic Verification of Sequential Consistency. In *Proceedings of the 32nd Annual International Symposium on Computer Architecture*, pp. 482–493, June 2005. doi:10.1109/ISCA.2005.25

[46] A. Meixner and D. J. Sorin. Dynamic Verification of Memory Consistency in Cache-Coherent Multithreaded Computer Architectures. In *Proceedings of the International Conference on Dependable Systems and Networks*, pp. 73–82, June 2006. doi:10.1109/DSN.2006.29

[47] A. Meixner and D. J. Sorin. Error Detection Using Dynamic Dataflow Verification. In *Proceedings of the International Conference on Parallel Architectures and Compilation Techniques*, pp. 104–115, Sept. 2007.

[48] A. Meixner and D. J. Sorin. Error Detection via Online Checking of Cache Coherence with Token Coherence Signatures. In *Proceedings of the Twelfth International Symposium on High-Performance Computer Architecture*, pp. 145–156, Feb. 2007.

[49] P. Montesinos, W. Liu, and J. Torrellas. Using Register Lifetime Predictions to Protect Register Files Against Soft Errors. In *Proceedings of the International Conference on Dependable Systems and Networks*, June 2007.

[50] S. S. Mukherjee, J. Emer, T. Fossum, and S. K. Reinhardt. Cache Scrubbing in Microprocessors: Myth or Necessity? In *10th IEEE Pacific Rim International Symposium on Dependable Computing (PRDC'04)*, pp. 37–42, Mar. 2004. doi:10.1109/PRDC.2004.1276550

[51] S. S. Mukherjee, M. Kontz, and S. K. Reinhardt. Detailed Design and Implementation of Redundant Multithreading Alternatives. In *Proceedings of the 29th Annual International Symposium on Computer Architecture*, pp. 99–110, May 2002.

[52] S. Narayanasamy, B. Carneal, and B. Calder. Patching Processor Design Errors. In *Proceedings of the International Conference on Computer Design*, Oct. 2006.

[53] M. Nicolaidis. Efficient Implementations of Self-Checking Adders and ALUs. In *Proceedings of the 23rd International Symposium on Fault-Tolerant Computing Systems*, pp. 586–595, June 1993. doi:10.1109/FTCS.1993.627361

[54] N. Oh, P. P. Shirvani, and E. J. McCluskey. Error Detection by Duplicated Instructions in Super-Scalar Processors. *IEEE Transactions on Reliability*, 51(1), pp. 63–74, Mar. 2002. doi:10.1109/24.994913

[55] A. Parashar, S. Gurumurthi, and A. Sivasubramaniam. SlicK: Slice-Based Locality Exploitation for Efficient Redundant Multithreading. In *Proceedings of the Twelfth International Conference on Architectural Support for Programming Languages and Operating Systems*, Oct. 2006.

[56] J. H. Patel and L. Y. Fung. Concurrent Error Detection in ALUs by Recomputing with Shifted Operands. *IEEE Transactions on Computers*, C-31(7), pp. 589–595, July 1982.

[57] K. Pattabiraman, G. P. Saggese, D. Chen, Z. Kalbarczyk, and R. K. Iyer. Dynamic Derivation of Application-Specific Error Detectors and Their Implementation in Hardware. In *Proceedings of the Sixth European Dependable Computing Conference*, 2006.

[58] P. Racunas, K. Constantinides, S. Manne, and S. S. Mukherjee. Perturbation-Based Fault Screening. In *Proceedings of the Twelfth International Symposium on High-Performance Computer Architecture*, pp. 169–180, Feb. 2007.

[59] V. K. Reddy and E. Rotenberg. Coverage of a Microarchitecture-level Fault Check Regimen in a Superscalar Processor. In *Proceedings of the International Conference on Dependable Systems and Networks*, June 2008.

[60] S. K. Reinhardt and S. S. Mukherjee. Transient Fault Detection via Simultaneous Multithreading. In *Proceedings of the 27th Annual International Symposium on Computer Architecture*, pp. 25–36, June 2000. doi:10.1145/339647.339652

[61] G. A. Reis, J. Chang, N. Vachharajani, R. Rangan, and D. I. August. SWIFT: Software Implemented Fault Tolerance. In *Proceedings of the International Symposium on Code Generation and Optimization*, pp. 243–254, Mar. 2005. doi:10.1109/CGO.2005.34

[62] E. Rotenberg. AR-SMT: A Microarchitectural Approach to Fault Tolerance in Microprocessors. In *Proceedings of the 29th International Symposium on Fault-Tolerant Computing Systems*, pp. 84–91, June 1999. doi:10.1109/FTCS.1999.781037

[63] N. N. Sadler and D. J. Sorin. Choosing an Error Protection Scheme for a Microprocessor's L1 Data Cache. In *Proceedings of the International Conference on Computer Design*, Oct. 2006.

[64] J. H. Saltzer, D. P. Reed, and D. D. Clark. End-to-End Arguments in Systems Design. *ACM Transactions on Computer Systems*, 2(4), pp. 277–288, Nov. 1984. doi:10.1145/357401.357402

[65] S. Sarangi, A. Tiwari, and J. Torrellas. Phoenix: Detecting and Recovering from Permanent Processor Design Bugs with Programmable Hardware. In *Proceedings of the 39th Annual IEEE/ACM International Symposium on Microarchitecture*, Dec. 2006.

[66] N. R. Saxena and E. J. McCluskey. Control-Flow Checking Using Watchdog Assists and Extended-Precision Checksums. *IEEE Transactions on Computers*, 39(4), pp. 554–559, Apr. 1990. doi:10.1109/12.54849

[67] E. Schuchman and T. N. Vijaykumar. BlackJack: Hard Error Detection with Redundant Threads on SMT. In *Proceedings of the International Conference on Dependable Systems and Networks*, pp. 327–337, June 2007.

[68] M. A. Schuette and J. P. Shen. Processor Control Flow Monitoring Using Signatured Instruction Streams. *IEEE Transactions on Computers*, C-36(3), pp. 264–276, Mar. 1987.

[69] F. F. Sellers, M.-Y. Hsiao, and L. W. Bearnson. *Error Detecting Logic for Digital Computers*. McGraw Hill Book Company, 1968.

[70] F. W. Shih. High Performance Self-Checking Adder for VLSI Processor. In *Proceedings of IEEE 1991 Custom Integrated Circuits Conference*, pp. 15.7.1–15.7.3, 1991. doi:10.1109/CICC.1991.164039

[71] P. Shivakumar, M. Kistler, S. W. Keckler, D. Burger, and L. Alvisi. Modeling the Effect of Technology Trends on the Soft Error Rate of Combinational Logic. In *Proceedings of the International Conference on Dependable Systems and Networks*, June 2002. doi:10.1109/DSN.2002.1028924

[72] S. Shyam, K. Constantinides, S. Phadke, V. Bertacco, and T. Austin. Ultra Low-Cost Defect Protection for Microprocessor Pipelines. In *Proceedings of the Twelfth International Conference on Architectural Support for Programming Languages and Operating Systems*, Oct. 2006. doi:10.1145/1168857.1168868

[73]   T. J. Slegel et al. IBM's S/390 G5 Microprocessor Design. *IEEE Micro*, pp. 12–23, Mar./Apr. 1999. doi:10.1109/40.755464

[74]   J. C. Smolens et al. Fingerprinting: Bounding the Soft-Error Detection Latency and Bandwidth. In *Proceedings of the Eleventh International Conference on Architectural Support for Programming Languages and Operating Systems*, Oct. 2004.

[75]   J. C. Smolens, B. T. Gold, B. Falsafi, and J. C. Hoe. Reunion: Complexity-Effective Multicore Redundancy. In *Proceedings of the 41st Annual IEEE/ACM International Symposium on Microarchitecture*, Nov. 2008.

[76]   J. C. Smolens, B. T. Gold, J. C. Hoe, B. Falsafi, and K. Mai. Detecting Emerging Wearout Faults. In *Proceedings of the Workshop on Silicon Errors in Logic—System Effects*, Apr. 2007.

[77]   J. C. Smolens, J. Kim, J. C. Hoe, and B. Falsafi. Efficient Resource Sharing in Concurrent Error Detecting Superscalar Microarchitectures. In *Proceedings of the 37th Annual IEEE/ACM International Symposium on Microarchitecture*, Dec. 2004. doi:10.1109/MICRO.2004.19

[78]   E. S. Sogomonyan, D. Marienfeld, V. Ocheretnij, and M. Gossel. A New Self-Checking Sum-Bit Duplicated Carry-Select Adder. In *Proceedings of the Design, Automation, and Test in Europe Conference*, 2004. doi:10.1109/DATE.2004.1269087

[79]   D. J. Sorin, M. D. Hill, and D. A. Wood. Dynamic Verification of End-to-End Multiprocessor Invariants. In *Proceedings of the International Conference on Dependable Systems and Networks*, pp. 281–290, June 2003. doi:10.1109/DSN.2003.1209938

[80]   J. Srinivasan, S. V. Adve, P. Bose, and J. A. Rivers. The Impact of Technology Scaling on Lifetime Reliability. In *Proceedings of the International Conference on Dependable Systems and Networks*, June 2004. doi:10.1109/DSN.2004.1311888

[81]   Sun Microsystems. UltraSPARC IV Processor Architecture Overview. Sun Microsystems Technical Whitepaper, Feb. 2004.

[82]   K. Sundaramoorthy, Z. Purser, and E. Rotenberg. Slipstream Processors: Improving Both Performance and Fault Tolerance. In *Proceedings of the Ninth International Conference on Architectural Support for Programming Languages and Operating Systems*, pp. 257–268, Nov. 2000.

[83]   W. J. Townsend, J. A. Abraham, and E. E. Swartzlander, Jr. Quadruple Time Redundancy Adders. In *Proceedings of the 18th IEEE International Symposium on Defect and Fault Tolerance in VLSI Systems*, pp. 250–256, Nov. 2003. doi:10.1109/DFTVS.2003.1250119

[84]   D. M. Tullsen, S. J. Eggers, J. S. Emer, H. M. Levy, J. L. Lo, and R. L. Stamm. Exploiting Choice: Instruction Fetch and Issue on an Implementable Simultaneous Multithreading Processor. In *Proceedings of the 23rd Annual International Symposium on Computer Architecture*, pp. 191–202, May 1996.

[85]   D. P. Vadusevan and P. K. Lala. A Technique for Modular Design of Self-Checking Carry-Select Adder. In *Proceedings of the 20th IEEE International Symposium on Defect and Fault Tolerance in VLSI Systems*, 2005.

[86] J. von Neumann. Probabilistic Logics and the Synthesis of Reliable Organisms from Unreliable Components. In C. E. Shannon and J. McCarthy, editors, *Automata Studies*, pp. 43–98. Princeton University Press, Princeton, NJ, 1956.

[87] I. Wagner, V. Bertacco, and T. Austin. Shielding Against Design Flaws with Field Repairable Control Logic. In *Proceedings of the Design Automation Conference*, July 2006. doi:10.1145/1146909.1146998

[88] J. F. Wakerly. *Error Detecting Codes, Self-Checking Circuits and Applications*. North Holland, 1978.

[89] N. J. Wang and S. J. Patel. ReStore: Symptom-Based Soft Error Detection in Microprocessors. *IEEE Transactions on Dependable and Secure Computing*, 3(3), pp. 188–201, 2006. doi:10.1109/TDSC.2006.40

[90] N. J. Warter and W.-M. W. Hwu. A Software Based Approach to Achieving Optimal Performance for Signature Control Flow Checking. In *Proceedings of the 20th International Symposium on Fault-Tolerant Computing Systems*, pp. 442–449, June 1990. doi:10.1109/FTCS.1990.89399

[91] C. Weaver and T. Austin. A Fault Tolerant Approach to Microprocessor Design. In *Proceedings of the International Conference on Dependable Systems and Networks*, pp. 411–420, July 2001. doi:10.1109/DSN.2001.941425

[92] K. Wilken and J. P. Shen. Continuous Signature Monitoring: Low-Cost Concurrent Detection of Processor Control Errors. *IEEE Transactions on Computer-Aided Design*, 9(6), pp. 629–641, June 1990. doi:10.1109/43.55193

[93] Y. C. Yeh. Triple-Triple Redundant 777 Primary Flight Computer. In *Proceedings of the Aerospace Applications Conference*, pp. 293–307, volume 1, Feb. 1996. doi:10.1109/AERO.1996.495891

[94] W. Zhang. Enhancing Data Cache Reliability by the Addition of a Small Fully-Associative Replication Cache. In *Proceedings of the 18th Annual International Conference on Supercomputing*, pp. 12–19, June 2004. doi:10.1145/1006209.1006212

[95] W. Zhang, S. Gurumurthi, M. Kandemir, and A. Sivasubramaniam. ICR: In-Cache Replication for Enhancing Data Cache Reliability. In *Proceedings of the International Conference on Dependable Systems and Networks*, pp. 291–300, June 2003.

[96] J. Ziegler et al. IBM Experiments in Soft Fails in Computer Electronics. *IBM Journal of Research and Development*, 40(1), pp. 3–18, Jan. 1996.

· · · ·

CHAPTER 3

Error Recovery

In Chapter 2, we learned how to detect errors. Detecting an error is sufficient for providing safety, but we would also like the system to recover from the error. Recovery hides the effects of the error from the user. After recovery, the system can resume operation and ideally remain live. For many systems, availability is the most important metric, and achieving high availability requires the system to be able to recover from its errors without user intervention. If the error was due to a permanent fault, recovery may not be sufficient for liveness because execution after recovery will keep reencountering the same permanent fault. The solutions to this problem—permanent fault diagnosis and self-repair—are the topics of the next two chapters.

In this chapter, we first discuss general concepts in error recovery (Section 3.1). We then present error recovery schemes that are specific to microprocessor cores (Section 3.2), caches and memory (Section 3.3), and multiprocessors (Section 3.4). We briefly discuss software-implemented error recovery (Section 3.5). We conclude with a discussion of open problems (Section 3.6).

3.1 GENERAL CONCEPTS

There are two primary approaches to error recovery. Forward error recovery (FER) corrects the error without reverting back to a previous state. An example of a FER scheme is triple modular redundancy (TMR) because the system continues to make forward progress in the presence of errors. The two correct modules outvote the module that suffers an error. Backward error recovery (BER) restores the state of the system to a known-good pre-error state. A common form of BER is to periodically checkpoint the state of the system and restore the system state to a pre-error checkpoint if an error is detected.

3.1.1 Forward Error Recovery

With FER, the system can correct the error in place and continue to make forward progress without restoring a prior state of the system. FER, like error detection, can be implemented using physical redundancy, information redundancy, or temporal redundancy. Fundamentally, FER requires more of each type of redundancy than error detection. If a given amount of redundancy is necessary to determine an error has occurred, then additional redundancy is required to correct that error.

Physical Redundancy. Recall from Chapter 2 that dual modular redundancy (DMR) is sufficient to detect errors. A mismatch between the results produced by the two replicas indicates an error. However, with just two replicas, error correction is impossible because the system cannot determine which replica produced the erroneous result. TMR provides the additional amount of redundancy, compared to DMR, that is required to correct a single error (i.e., errors in a single module). Naively extending this pattern, one might expect that 4-MR provides even better error correction, but the problem with 4-MR is that double errors are often still uncorrectable. Because of the possibility of "ties," where half the modules have the correct result and the other half have the same incorrect result, N-modular redundancy (NMR) schemes almost invariably choose an odd value for N. Because of the high hardware, power, and energy costs of NMR (roughly 200% for TMR), discussed in Chapter 2, it is a viable error recovery scheme only for small modules or mission-critical systems.

Information Redundancy. An error-correcting code (ECC) can provide FER. If a datum incurs an error while residing in ECC-protected memory, for example, then the ECC on the datum can be used to correct the error and provide the error-free datum. The Hamming distance (HD) of an error code determines how many bit errors in a word it can detect and correct. Recall from Chapter 2 that an HD enables the detection of HD−1 bit errors and the correction of (HD−1)/2 bit errors. A greater Hamming distance is required for correction than detection and thus more redundant bits are required to achieve correction than detection. The computations involved in ECC are also more complicated and require more time than the computations required for EDC.

Temporal Redundancy. To achieve FER, a temporal redundancy scheme needs to perform a given operation at least three times. If the operation is performed only twice, then a difference in the results indicates an error but does not enable the system to identify which of the two operations was correct. Performing the operation three times, analogously to TMR, enables the system to vote among the three results and correct a single erroneous result. Because of the performance impact of performing each operation at least three times, temporal redundancy is not used as often as physical or information redundancy for FER. FER with temporal redundancy also incurs a significant 200% energy overhead.

3.1.2   Backward Error Recovery

BER involves restoring the state of the system to a previous, known-good state of the system, called the *recovery point* (or *recovery line* for a system with multiple cores). Implementing BER requires an architect to answer six questions:

1. What state must be saved for the recovery point?
2. Which algorithm should be used for saving the recovery point?

3. Where should the recovery point be saved?

4. How should the recovery point state be restored during a recovery?

5. When can a recovery point be deallocated?

6. What does the system do after the recovery point state has been restored?

In this section, we focus on hardware-implemented BER, but we also mention several applications of software-implemented BER, including its extensive use in database management systems [9]. Before we discuss the six questions that BER designers must answer for a given system, there is one aspect of BER that applies to all systems that use BER: the *output commit problem* [5].

The Output Commit Problem. The output commit problem is that a system cannot communicate data to the "outside world" until it knows that these data are error-free. The outside world is anything that cannot be recovered with the BER scheme. Thus, errors must be contained within the sphere of recoverability so that the error does not propagate to a component that cannot be recovered. If an error escapes the sphere of recoverability, then the error is unrecoverable and the system fails. If a system with BER sends data to the outside world at time T and later detects an error and wishes to recover to a recovery point from before time T, it cannot undo having sent the data.

There are several options for choosing the sphere of recoverability, and the options are discussed at length by Gold et al. [8]. If BER is implemented just on the core, then errors cannot be allowed to propagate to the caches or memory or beyond. If BER includes the memory hierarchy, then errors can be allowed to propagate into the memory system but not to I/O devices. An example of a component that is outside the sphere of recoverability of any system is the printer. Once a request has been made to the printer and the printer has printed the document, it is generally impossible to undo the printing of the document even if the system subsequently discovers that the request to the printer was erroneous.

The common approach to the output commit problem is to wait to send data to the outside world until the error detection mechanisms have completed their checking of all operations before the sending of the data. Thus, the output commit problem places error detection on the critical path and degrades error-free performance. In the absence of output operations, BER schemes can usually hide most or all of the latency of error detection. Consider a system that saves a recovery point that reflects the state of the system at time T. If an error occurs at time $T+e$ and is detected at time $T+e+d$ (d is the detection latency), then the system can still recover to the recovery point at time T. The error detection latency, d, does not hurt performance in the error-free scenario.

The output commit problem is a fundamental issue for BER schemes. Some research, including ReViveI/O [21], has mitigated its impact by leveraging the semantics of specific devices in the outside world. For example, if we know that an operation is idempotent, such as a write to a given location on a disk, then we can perform the operation before we are certain it is error-free. If the system recovers to a state before this operation was performed, then performing it again is fine.

What State Must Be Saved for the Recovery Point. BER must recover the system to *a consistent, pre-error state from which it can resume execution.* For a processor to resume execution, it requires all of the architectural state, including the program counter, architectural registers, status registers, and the memory state. Furthermore, this architectural state must represent a *precise* [29] state of the processor. A precise state of a processor is one that (a) includes all of the effects of all instructions prior in program order to and including a given instruction and (b) does not include any state of any instructions that are after that instruction in program order.

There are two important issues in considering what state must be saved. First, there is no need to save microarchitectural state, such as the state of the branch predictor or the load-store queue. By saving the precise architectural state, we do not need any microarchitectural state. Although there is no need to save microarchitectural state, an architect could still choose to do so to speed up the execution after recovery.

Second, a BER scheme does not need to save the exact state of the processor; it only needs to save a consistent state. A clear example of this subtle difference is the memory system. The BER scheme must save the state of the memory system. Assume that block B has value 3 in the L1 data cache. A BER scheme could remember that B has value 3 in the cache, or it could instead remember that B has value 3 in memory and that it must invalidate B from the cache during recovery. Whether B gets restored into the cache or the memory after recovery does not matter.

Which Algorithm to Use for Saving the Recovery Point. There are many possible algorithms for saving the state of the recovery point. In this section, we discuss the two most important aspects of these algorithms. First, does the algorithm use checkpointing, logging, or a combination of the two? Second, for multiprocessors, how does the algorithm establish a consistent recovery line through the recovery points of all of the cores?

1. Checkpointing and logging: Checkpointing and logging are two mechanisms that provide the same functionality, but they have different advantages and disadvantages.

   With checkpointing, the processor decides, at certain times, to save its entire state. Checkpoints can be taken at regular periodic intervals or in response to certain events. Taking checkpoints more frequently is likely to increase the performance penalty of checkpointing, but it reduces the amount of error-free work that must be replayed after a recovery. For example, consider a processor that checkpoints itself every minute. If a failure occurs 59 seconds after the most recent checkpoint, all of the error-free work that occurred during those 59 seconds between the checkpoint and the error is lost. Checkpointing is useful in many contexts, not just for improving a processor's fault tolerance. For example, checkpointing a thread enables it to be restarted on another core for purposes of load balancing across cores. Software-implemented checkpointing is useful in many situations, including taking nightly snapshots of a file system.

With logging, a BER scheme records the changes that are made to the system state. Each log entry is logically a tuple <name, old value>. These logs of changes can be unrolled if an error is detected. Logging, like checkpointing, is useful in contexts other than architectural BER. Many programs, such as word processors and spreadsheets, log changes to data structures so that they can provide "Undo" functionality. Many operating systems log events that occur and then these logs can be mined to look for anomalies, such as those due to security breaches.

Because checkpointing and logging have different costs for different types of state, many BER systems use a hybrid of both. For example, SafetyNet [30] uses checkpointing to save the core's register state and it uses logging to save changes made to memory state.

2. Creating consistent multiprocessor recovery points. In a system with multiple cores, it can be challenging to create a *consistent* recovery line across all of the cores. For a recovery line to be consistent, it must respect causality; that is, the recovery line cannot include the effects of an event that has not occurred yet. The canonical example of an inconsistent recovery line is one that includes the effects of a message being received but not of that message being sent.

The challenge for creating consistent recovery lines is saving the state of communication between cores. In a multicore processor, it is not sufficient to independently save the state of each core; we must consider the state of the communication between the cores. Depending on the architecture, this communication state may include cache coherence state and the state of received or in-flight messages. In Figure 3.1, we illustrate the execution of a two-core system in which core 1 is sending messages to core 2. We show three possible recovery

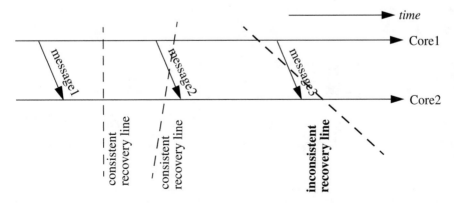

FIGURE 3.1: Examples of consistent and inconsistent multicore recovery lines. A consistent recovery line cannot include the reception of a message that has not yet been sent.

lines. Two of them are consistent, but the rightmost recovery line is inconsistent because it includes the reception of message3 by core 2 but not the sending of message3 by core 1.

There are two approaches to creating consistent recovery lines: uncoordinated and coordinated saving of recovery points.

With uncoordinated checkpointing (or logging), each core saves its own recovery point without coordinating with the others. The recovery line is the collection of individual recovery points. This uncoordinated option is simple to implement and it is fast in the common, error-free case. The problem is that, if an error is detected, having each core recover to its most recent recovery point may lead to an inconsistent recovery line. In Figure 3.2, when core 3 detects an error, it recovers to recovery point 3.3 (denoted RP 3.3). However, if core 3 reverts to RP 3.3, then the system is in a state in which core 2 has received msg7 but core 3 has not sent it yet. To remedy this issue, core 2 must revert to RP 2.3. However, this recovery leads to a state in which core 1 has received msg8 but core 2 has not sent it. Core 1 must now revert to RP 1.3. This recovery leads to core 3 having received msg6 before it was sent by core 1. This unraveling of recovery points does not lead to a consistent recovery line until all three cores are back to their original recovery points. That is, the only consistent recovery line is the collection of RP 1.1, 2.1, and 3.1. This pathological unraveling is called "cascading rollbacks" or the "domino effect," and it is the major drawback to uncoordinated saving of recovery points.

The natural alternative to uncoordinated saving of recovery points is to have the cores coordinate among themselves to save a consistent recovery line. A core or central controller can initiate the procedure of saving the recovery line. The simplest option is a procedure in which all of the cores wait for all in-flight messages to arrive at their destinations and then, when the system has quiesced, each core saves its own local recovery point. The collection of recovery points is consistent because

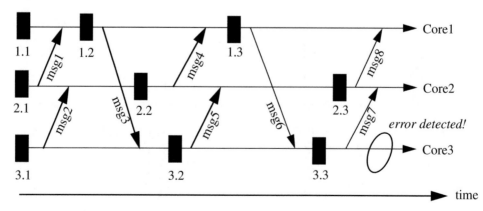

FIGURE 3.2: Example of cascading rollbacks (the "domino effect").

there are no in-flight communication. There are other algorithms that are more aggressive and offer better performance, and we discuss one of them in Section 3.4.

Where to Save the Recovery Point. For the recovery point state to be useful, it must be protected from errors. Most software-implemented BER schemes, such as those for database management systems, save their recovery point state on disk, and they assume that disks are stable storage. This assumption is generally valid because of the ECC on disks, and disks can be made even more trustworthy by protecting them with RAID [22]. Hardware-implemented BER schemes generally save data on disks or main memory. Some hardware BER schemes use caches and on-chip shadow register files for saving recovery point state.

How to Restore the Recovery Point State. There are two issues involved in restoring the recovery point state. First, the system must be careful to flush out all potentially corrupted state. Second, if the system has multiple options for where to put the recovery point state (e.g., in cache or in memory), it must decide which option is appropriate.

When to Deallocate a Recovery Point. The current recovery point state cannot be deallocated until another more recent recovery point has been successfully saved. Otherwise, a detected error would be unrecoverable because there would be no recovery point. Saved state from before the most recent recovery point can be discarded because, in the case of a detected error, the system would revert to the most recent recovery point instead of needlessly reverting to an even older state.

A key issue in deallocation is when a checkpoint (for brevity, we use the term *checkpoint* in this discussion, instead of considering both checkpoints and logs) is validated as being error-free. Until this point, the error detection mechanisms are still determining whether the checkpoint is error-free. One consequence of error detection latency is that it impacts how long a checkpoint must be kept until it can be designated the recovery point. Long error detection latencies thus often motivate the pipelining of checkpoints, as illustrated in Figure 3.3. In this figure, there is a single recovery point and multiple more recent checkpoints that have not yet been validated as error-free.

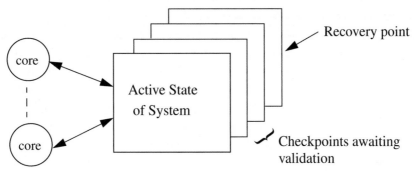

FIGURE 3.3: Pipelined checkpoints.

When the oldest nonvalidated checkpoint is determined to be error-free, it becomes the new recovery point and the old recovery point is deallocated. The advantage of pipelining is that it can take error detection latency off the critical path. Consider a system with just a single checkpoint that is the recovery point. To create a new recovery point, the system's normal execution stops and the system must wait for the error detection mechanisms to validate the currently active state and then save it as the new recovery point. With pipelining, this error detection can be performed in parallel with normal execution. The primary cost of pipelined checkpointing is the hardware cost of the additional storage to hold the nonvalidated checkpoints.

Because the issue of recovery point deallocation depends on the error detection mechanisms, rather than the BER scheme itself, we do not discuss it again when we present BER for specific processor components later in this chapter.

What to Do After Recovery. After recovering to the recovery point, most systems just try to resume execution from that point. If the system executes past where the recovery-triggering error occurred previously, the system can assume the error was transient. However, if the system encounters the same error again, the error is likely due to a permanent fault or a design bug. In either of these situations, the system cannot continue to make forward progress. In Chapter 5, we discuss how a processor can repair itself in these situations so as to make forward progress. We do not discuss this issue again in this chapter.

3.1.3   Comparing the Performance of FER and BER

The relative performances of FER and BER depend on several factors. We summarize the performance issues in Table 3.1 and discuss them next.

TABLE 3.1: FER versus BER Performance.		
	FER	**BER**
Error detection	On critical path	Off critical path (if no output)
Error-free performance penalty	Small/medium: due to error detection latency	Small: due to saving state (may be worse if frequent output)
Penalty when error occurs	Small: latency to correct error	Medium/large: latency to restore state and replay lost work

Forward Error Recovery. During error-free execution, most FER schemes incur a slight performance penalty for error detection. Because FER schemes cannot recover to a prior state, they cannot commit any operation until it has been determined to be error-free. Effectively, for systems with FER, *all* operations are output operations and are subject to the output commit problem. Thus, error detection is on the critical path for FER. When an error occurs, FER incurs little additional performance penalty to correct it.

Backward Error Recovery. During error-free execution, most BER schemes incur a slight performance penalty for saving state. This penalty is a function of how often state is saved and how long it takes to save it. In the absence of output operations, BER schemes can often take error detection off the critical path because, even if an error is detected after the erroneous operation has been allowed to proceed, the processor can still recover to a pre-error checkpoint. To overlap the latency of error detection requires pipelined checkpointing, as described in "When to Deallocate a Recovery Point" from Section 3.1.2. When an error occurs, BER incurs a relatively large penalty to restore the recovery point and replay the work since the recovery point that was lost.

3.2 MICROPROCESSOR CORES
Both FER and BER approaches exist for microprocessor cores.

3.2.1 FER for Cores
The only common FER scheme for an entire core is TMR. With three cores and a voter, an error in a single core is corrected when the result of that core is outvoted by the other two cores.

Within a core, TMR can be applied to specific units, although this is rare in commodity cores due to the hardware and power costs for TMR.

A more common approach for FER within a core is the use of ECC. By protecting storage (e.g., register file) or a bus with ECC, the core can correct errors without needing to restore a previous state. However, even ECC may be infeasible in many situations because it is on the critical path, and high-performance cores often have tight timing constraints.

3.2.2 BER for Cores
BER for cores is a well-studied issue because of the long history of checkpoint/recovery hardware for commercial cores. IBM has long incorporated checkpoint/recovery into its mainframe processor cores [28]. Outside of mainframe processor cores, checkpoint/recovery hardware often exists, but it is used for recovering from the effects of misspeculation instead of being used for error recovery. A core that speculatively executes instructions based on a branch prediction may later discover that the prediction was incorrect. To hide the effects of the misspeculated instructions from the software,

the core recovers to a pre-speculation checkpoint and resumes execution down the correct control flow path. In this situation, a misprediction is analogous to an error. In both situations, subsequent instructions are executed erroneously and their effects must be undone.

With little additional effort, the existing checkpoint/recovery mechanisms used for supporting speculation can be used for error recovery. However, there are two important aspects of error recovery that differ. First, for error recovery purposes, a core would likely take less frequent checkpoints (or log less frequently). Errors are less likely than mis-speculations, and thus the likelihood of losing the work done between a checkpoint and when an error is detected is far less than the likelihood of losing the work done between a checkpoint and when a mis-prediction is detected. Second, for error recovery purposes, we may wish to protect the recovery point state from errors. This protection is not required for speculation purposes that assume that errors do not occur.

Design Options. There are several points to consider in implementing BER.

1. What state to save for the recovery point. Implementing BER for a core is fairly simple because there is a relatively small amount of architectural state that must be saved. This state includes the general purpose registers and the other architecturally visible registers, including core status registers (e.g., processor status word). We defer the discussion of memory state until Section 3.3; for now, assume the core performs no stores.

2. Which algorithm to use for saving the recovery point. Cores can use either checkpointing or logging to save state, and both algorithms have been used in practice. The choice of algorithm often depends on the exact microarchitecture of the core and the granularity of recovery that is desired. If there are few registers and recoveries are infrequent, then checkpointing is probably preferable. If there are many registers and recoveries are frequent, then logging is perhaps a better option.

3. Where to save the recovery point. Virtually all cores save their state in structures within the core. Using a shadow register file or register renaming table is a common approach. The only schemes that save this state off-chip are those using BER for highly reliable systems rather than for supporting speculative execution. To avoid the possibility of a corrupted recovery point, which would make recovery impossible, an architect may wish to add ECC to the recovery point state.

4. How to restore the recovery point state. Before copying the core's recovery point back into its operational registers, we must flush all of the core's microarchitectural state, such as the reorder buffer, reservation stations, and load-store queue. These microarchitectural structures may hold state related to instructions that were squashed during recovery, and we need to remove this state from the system.

Recent Developments in Core BER. Checkpoint/recovery hardware has recently enjoyed a resurgence in cores for a variety of reasons.

1. Error recovery. The cores in IBM's zSeries systems have long had checkpoint/recovery hardware [20]. Recently, though, IBM has extended checkpoint/recovery to its POWER6 microarchitecture [17] that it uses in its Power^{TM} Systems.

2. Transactional memory. There has been a recent surge of interest in using transactional memory [10] as a programming paradigm for multicore processors. Architects have begun adding hardware support for transactional memory, and one useful feature is the ability to recover a core that is executing a transaction that is discovered to conflict with another transaction. Sun's Rock processor has added checkpoint/recovery [18]. Software running on Rock can invoke an instruction that causes the core to save its register state in a set of shadow registers.

3. Scalable core design. Akkary et al. [2] observed that superscalar cores could be made more scalable—that is, able to extract more instruction level parallelism—using checkpointing to implement larger instruction windows. Because this topic is outside the scope of fault tolerance, we mention it only to show the possible synergy between BER and checkpoint/recovery for other purposes.

3.3 SINGLE-CORE MEMORY SYSTEMS

In Section 3.2, we discussed error recovery for cores without considering the memory system. This is an unrealistic assumption because all cores interact with various memory structures, including caches, memory, and translation lookaside buffers (TLBs). In this section, we consider memory systems for single-core processors. In Section 3.4, we address error recovery issues that are specific to multicore processors, including shared memory systems.

3.3.1 FER for Caches and Memory

The only commonly used FER scheme for memory structures is ECC. Other possible FER schemes, such as providing three or more replicas of an item in a memory structure, are prohibitively expensive.

ECC can be used at many different granularities, including word and block. The area overhead of using ECC can be decreased by applying it at a coarser granularity; a coarse granularity complicates accesses to data that are smaller than the ECC granularity.

One interesting twist on ECC is RAID-M or Chipkill Memory [3, 12]. As both of its commonly used names imply, the idea is to use a RAID [22]-like approach to recover from errors that permanently kill memory (DRAM) chips. This chipkill error model reflects the many possible underlying physical phenomena that can cause an entire DRAM chip to fail. Implementations of RAID-M include one or more extra DRAM chips, and the data are spread across the original and redundant chips such that the system can recover from the loss of all of the data on any single chip.

3.3.2   BER for Caches and Memory

As was the case for microprocessor cores, the use of hardware to enable recovery for caches and memory has increased recently, and the reasons for this increased use are the same. In addition to providing error recovery, the goals are to support speculation, large instruction windows, and transactional memory. Being able to recover just the core is insufficient, unless the core is restricted from committing stores. Throttling stores "solves" the problem, but throttling also limits the amount of speculation that can be performed or instruction-level parallelism that can be exploited. To overcome this limitation, stores must be allowed to modify memory state and we must add some mechanism for recovering that memory state in the case of an error (or misprediction).

What State to Save for Recovery Point. The architectural state of the memory system includes the most recent values of every memory address. If the only copy of the most recent value for a memory address is in a cache, then that value must be saved. Although TLBs are caches, they never hold the only copy of the most recent value for a memory address. TLBs hold only read-only copies of data that are also in memory.

Which Algorithm to Use for Saving the Recovery Point. Because the size of memory is usually immense, a pure checkpointing scheme, like those often used for core register files, is prohibitively expensive. Copying the entire memory image would require a large amount of time and extra storage. Instead, logging changes made to memory values is likely to be far more efficient. An example of a logging scheme is SafetyNet [30], which creates logical checkpoints using logging. After a new checkpoint is logically created, SafetyNet logs the old value of any memory location that is overwritten. Because recoveries are only performed at the granularity of checkpoints, rather than to arbitrary points within a log, SafetyNet logs only the first write of each memory location between checkpoints; once the value of a location from the time of the checkpoint has been logged, additional logging of that location is unnecessary. The recovery process consists of walking backward through the log to restore the values that existed at the time of the checkpoint's creation. In Figure 3.4, we illustrate an example of using logging to implement logical checkpointing, similar to the SafetyNet [30] approach.

Where to Save the Recovery Point. The decision where to save the recovery point state depends greatly on the purpose of the recovery scheme. For some core speculation approaches, a large, perhaps multilevel store queue may be sufficient to hold the values of stores that may need to be undone. For longer periods of speculation, architects have proposed adding buffers to hold the state

```
// Assume all memory locations are initially zero
// Assume checkpoint taken now before this snippet of code
store 3, Mem[0]          // log that Mem[0] was 0 at checkpoint
store 4, Mem[1]          // log that Mem[1] was 0 at checkpoint
store 5, Mem[0]          // do not need to log Mem[0] again
store 6, Mem[2]          // log that Mem[2] was 0 at checkpoint
// Undoing the log would put the value zero in memory locations 0, 1, and 2
```

FIGURE 3.4: Example of using logging to implement logical checkpointing of memory.

of committed stores [7, 26]. These buffers effectively serve as logs. For purposes of fault tolerance, we must trade off our wish to keep the data in the safest place versus our wish to keep the performance overhead low. Generally, this trade-off has led to saving the recovery point state in caches and memory rather than in the safer but vastly slower disk.

One of the landmark papers on BER, Hunt and Marinos's [11] Cache-Aided Rollback Error Recovery (CARER) explores how to use the cache to hold recovery point state. CARER permits committed stores to write into the cache, but it does not allow them to be written back to memory until they have been validated as being error-free. Thus, the memory and the clean lines in the cache represent the recovery point state. Dirty lines in the cache represent state that could be recovered if an error is detected. During a recovery, all dirty lines in the cache are invalidated. If the address of one of these lines is accessed after recovery, it will miss in the cache and obtain the recovery point value for that data from memory.

How to Restore the Recovery Point State. Any cache or memory state, including TLB entries, that is not part of the recovery point, should be flushed. Otherwise, we risk keeping state that was generated by instructions that executed after the recovery point.

A key observation made in the CARER paper is that the memory state does not need to be restored to the same place where it had been. For example, assume that data block B had been in the data cache with the value 3 when the checkpoint was taken. The recovery process could restore block B to the value 3 in either the data cache or the memory. These placements of the restored data are architecturally equivalent.

3.4   ISSUES UNIQUE TO MULTIPROCESSORS

BER for multiprocessors, including multicore processors, has one major additional aspect: how to handle the state of communication between cores. Depending on the architecture, this communication state may include cache coherence state and the state of received or in-flight messages. We focus here on cache-coherent shared memory systems because of their prevalence. We refer readers interested in BER for message-passing architectures to the excellent survey paper on that topic by Elnozahy et al. [4].

3.4.1   What State to Save for the Recovery Point

The architectural state of a multiprocessor includes the state of the cores, caches, and memories, plus the communication state. For the cache-coherent shared memory systems that we focus on in this discussion, the communication state may include cache coherence state. To illustrate why we may need to save coherence state, consider the following example for a two-core processor that uses its caches to save part of its recovery point state (like CARER [11]). When the recovery point is saved, core 1 has block B in a modified (read–write) coherence state, and core 2's cached copy of block B is invalid (not readable or writeable). If, after recovery, the coherence state is not restored properly, then both core 1 and core 2 may end up having block B in the modified state and thus both might believe they can write to block B. Having multiple simultaneous writers violates the single-writer/multiple-reader invariant maintained by coherence protocols and is likely to lead to a coherence violation.

3.4.2   Which Algorithm to Use for Saving the Recovery Point

As we discussed in "When to Deallocate a Recovery Point" from Section 3.1.2 the key challenge for saving the recovery line of a multicore processor is saving a *consistent* state of the system. We must save a recovery line from which the entire system can recover. The BER algorithm must consider how to create a consistent checkpoint despite the possibility of in-flight communication, such as a message currently in transit from core 1 to core 2. Uncoordinated checkpointing suffers from the cascading rollback problem described in "Which Algorithm to Use for Saving the Recovery Point" from Section 3.1.2, and thus we consider only coordinated checkpointing schemes now.

The simplest coordinated checkpointing solution is to quiesce the system and let all in-flight messages arrive at their destinations. Once there are no messages in-flight, the system establishes a recovery line by having each core save its own recovery point (including its caches and memory). This collection of core checkpoints represents a consistent system-wide recovery point. Quiescing the system is a simple and easy-to-implement solution, and it was used by a multiprocessor extension to CARER [1]. More recently, the ReVive BER scheme [25] used the quiescing approach, but for a wider range of system architectures. CARER is limited to snooping coherence, and ReVive considers modern multiprocessors with directory-based coherence. The drawback to this simple quiescing approach is the performance loss incurred while waiting for in-flight messages to arrive at their destinations.

To avoid the performance degradation associated with quiescing the system, SafetyNet [30] takes coordinated, pipelined checkpoints that are consistent in *logical time* [15] instead of physical time. Logical time is a time base that respects causality, and it has long been used to coordinate events in distributed systems. In the SafetyNet scheme, each core takes a checkpoint at the same logical time, without quiescing the system. The problem of in-flight messages is eliminated by checkpointing their effects in logical time.

One possible optimization for creating consistent checkpoints is to reduce the number of cores that must participate. For example, if core 1 knows that it has not interacted with any other cores since the last consistent checkpoint was taken, it does not have to take a checkpoint when the other cores decide to do so. If an error is detected and the system decides to recover, core 1 can recover to its older recovery point. The collection of core 1's older recovery point and the newer recovery points of the other cores represents a consistent system-wide recovery line. To exploit this opportunity, each core must track its interactions with the other cores. This optimization has been explored by the multiprocessor CARER [1], as well as in other work [34].

3.4.3 Where to Save the Recovery Point

The simplest option for saving coherence state is to save it alongside the values in the cache. If caches are used to hold recovery point state, then coherence state can be saved alongside the corresponding data in the cache. If the caches are not used to hold recovery point state, then coherence state does not need to be saved.

3.4.4 How to Restore the Recovery Point State

This issue has no multiprocessor-specific aspects.

3.5   SOFTWARE-IMPLEMENTED BER

Software BER schemes have been developed at radically different engineering costs from hardware BER schemes. Because software BER is a large field and not the primary focus of this book, we provide a few highlights from this field rather than an extensive discussion.

A wide range of systems use software BER. Tandem machines before the S2 (e.g., the Tandem NonStop) use a checkpointing scheme in which every process periodically checkpoints its state on another processor [27]. If a processor fails, its processes are restarted on the other processors that hold the checkpoints. Condor [16], a batch job management tool, can checkpoint jobs to restart them on other machines. Applications need to be linked with the Condor libraries so that Condor can checkpoint them and restart them. Other schemes, including work by Plank [23, 24] and Wang and Hwang [32, 33], use software to periodically checkpoint applications for purposes of fault tolerance. These schemes differ from each other primarily in the degree of support required from the programmer, linked libraries, and the operating system.

IEEE's Scalable Coherent Interface (SCI) standard specifies software support for BER [13]. SCI can perform end-to-end error retry on coherent memory transactions, although the specification describes error recovery as being "relatively inefficient." Recovery is further complicated for SCI accesses to its noncoherent control and status registers because some of these actions may have side effects.

Software BER schemes have also been developed for use in systems with software distributed shared memory (DSM). Software DSM, as the name suggests, is a software implementation of shared memory. Sultan et al. [31] developed a fault tolerance scheme for a software DSM scheme with the home-based lazy release consistency memory model. Wu and Fuchs [35] used a twin-page disk storage system to perform user-transparent checkpoint/recovery. At any point in time, one of the two disk pages is the working copy and the other page is the checkpoint. Similarly, Kim and Vaidya [14] developed a scheme that ensures that there are at least two copies of a page in the system. Morin et al. [19] leveraged a Cache Only Memory Architecture to ensure that at least two copies of a block exist at all times; traditional COMA schemes ensure the existence of only one copy. Feeley et al. [6] implemented log-based coherence for a transactional DSM.

3.6   CONCLUSIONS

Error recovery is a well-studied field with a wide variety of good solutions. Applying these solutions to new systems requires good engineering, but we do not believe there are as many interesting open problems in this field as there are in the other three aspects of fault tolerance. In addition to improving implementations, particularly for many-core processors, we believe architects will address two promising areas:

- Mitigating the output commit problem: The output commit problem is a fundamental limitation for BER schemes. Some research has explored techniques that leverage the semantics of specific output devices to hide the performance penalty of the output commit problem [21]. Another possible approach is to extend the processor's sphere of recoverability to reduce the size of the outside world. If architects can obtain access to devices that are currently unrecoverable—and are thus part of the outside world—then they can devise BER schemes that include these devices. Such research would involve a significant change in interfaces and may be too disruptive, but it could mitigate the impact of the output commit problem.
- Unifying BER for multiple purposes: We discussed how BER is useful for many purposes, not just in fault tolerance. There are opportunities to use a single BER mechanism to simultaneously support several of these purposes, and architects may wish to delve into BER implementations that can efficiently satisfy the demands of these multiple purposes.

3.7 REFERENCES

[1] R. E. Ahmed, R. C. Frazier, and P. N. Marinos. Cache-Aided Rollback Error Recovery (CARER) Algorithms for Shared-Memory Multiprocessor Systems. In *Proceedings of the 20th International Symposium on Fault-Tolerant Computing Systems*, pp. 82–88, June 1990. doi:10.1109/FTCS.1990.89338

[2] H. Akkary, R. Rajwar, and S. T. Srinivasan. Checkpoint Processing and Recovery: Towards Scalable Large Instruction Window Processors. In *Proceedings of the 36th Annual IEEE/ACM International Symposium on Microarchitecture*, Dec. 2003. doi:10.1109/MICRO.2003.1253246

[3] T. J. Dell. A White Paper on the Benefits of Chipkill-Correct ECC for PC Server Main Memory. IBM Microelectronics Division Whitepaper, Nov. 1997.

[4] E. Elnozahy, D. Johnson, and Y. Wang. A Survey of Rollback-Recovery Protocols in Message-Passing Systems. Technical Report CMU-CS-96-181, Department of Computer Science, Carnegie Mellon University, Sept. 1996.

[5] E. Elnozahy and W. Zwaenepoel. Manetho: Transparent Rollback-Recovery with Low Overhead, Limited Rollback, and Fast Output Commit. *IEEE Transactions on Computers*, 41(5), pp. 526–531, May 1992. doi:10.1109/12.142678

[6] M. Feeley, J. Chase, V. Narasayya, and H. Levy. Integrating Coherency and Recoverability in Distributed Systems. In *Proceedings of the First USENIX Symposium on Operating Systems Design and Implementation*, pp. 215–227, Nov. 1994.

[7] C. Gniady, B. Falsafi, and T. Vijaykumar. Is SC + ILP = RC? In *Proceedings of the 26th Annual International Symposium on Computer Architecture*, pp. 162–171, May 1999. doi:10.1145/307338.300993

[8] B. T. Gold, J. C. Smolens, B. Falsafi, and J. C. Hoe. The Granularity of Soft-Error Containment in Shared Memory Multiprocessors. In *Proceedings of the Workshop on System Effects of Logic Soft Errors*, Apr. 2006.

[9] J. Gray and A. Reuter. *Transaction Processing: Concepts and Techniques*. Morgan Kaufmann Publishers, 1993.

[10] M. Herlihy and J. E. B. Moss. Transactional Memory: Architectural Support for Lock-Free Data Structures. In *Proceedings of the 20th Annual International Symposium on Computer Architecture*, pp. 289–300, May 1993. doi:10.1109/ISCA.1993.698569

[11] D. Hunt and P. Marinos. A General Purpose Cache-Aided Rollback Error Recovery (CARER) Technique. In *Proceedings of the 17th International Symposium on Fault-Tolerant Computing Systems*, pp. 170–175, 1987.

[12] IBM. Enhancing IBM Netfinity Server Reliability: IBM Chipkill Memory. IBM Whitepaper, Feb. 1999.

[13] IEEE Computer Society. *IEEE Standard for Scalable Coherent Interface (SCI)*, Aug. 1993.

[14] J.-H. Kim and N. Vaidya. Recoverable Distributed Shared Memory Using the Competitive Update Protocol. In *Pacific Rim International Symposium on Fault-Tolerant Systems*, Dec. 1995.

[15] L. Lamport. Time, Clocks and the Ordering of Events in a Distributed System. *Communications of the ACM*, 21(7), pp. 558–565, July 1978. doi:10.1145/359545.359563

[16] M. Litzkow, T. Tannenbaum, J. Basney, and M. Livny. Checkpoint and Migration of UNIX Processes in the Condor Distributed Processing System. Technical Report 1346, Computer Sciences Department, University of Wisconsin–Madison, Apr. 1997.

[17] M. J. Mack, W. M. Sauer, S. B. Swaney, and B. G. Mealey. IBM POWER6 Reliability. *IBM Journal of Research and Development*, 51(6), pp. 763–774, 2007.

[18] M. Moir, K. Moore, and D. Nussbaum. The Adaptive Transactional Memory Test Platform: A Tool for Experimenting with Transactional Code for Rock. In *Proceedings of the 3rd ACM SIGPLAN Workshop on Transactional Computing*, Feb. 2008.

[19] C. Morin, A. Gefflaut, M. Banatre, and A.-M. Kermarrec. COMA: An Opportunity for Building Fault-Tolerant Scalable Shared Memory Multiprocessors. In *Proceedings of the 23rd Annual International Symposium on Computer Architecture*, pp. 56–65, May 1996.

[20] M. Mueller, L. Alves, W. Fischer, M. Fair, and I. Modi. RAS Strategy for IBM S/390 G5 and G6. *IBM Journal of Research and Development*, 43(5/6), Sept./Nov. 1999.

[21] J. Nakano, P. Montesinos, K. Gharachorloo, and J. Torrellas. ReViveI/O: Efficient Handling of I/O in Highly-Available Rollback-Recovery Servers. In *Proceedings of the Twelfth International Symposium on High-Performance Computer Architecture*, pp. 200–211, Feb. 2006.

[22] D. A. Patterson, G. Gibson, and R. H. Katz. A Case for Redundant Arrays of Inexpensive Disks (RAID). In *Proceedings of 1988 ACM SIGMOD Conference*, pp. 109–116, June 1988. doi:10.1145/50202.50214

[23] J. S. Plank. An Overview of Checkpointing in Uniprocessor and Distributed Systems, Focusing on Implementation and Performance. Technical Report UT-CS-97-372, Department of Computer Science, University of Tennessee, July 1997.

[24] J. S. Plank, K. Li, and M. A. Puening. Diskless Checkpointing. *IEEE Transactions on Parallel and Distributed Systems*, 9(10), pp. 972–986, Oct. 1998. doi:10.1109/71.730527

[25] M. Prvulovic, Z. Zhang, and J. Torrellas. ReVive: Cost-Effective Architectural Support for Rollback Recovery in Shared-Memory Multiprocessors. In *Proceedings of the 29th Annual International Symposium on Computer Architecture*, pp. 111–122, May 2002. doi:10.1109/ISCA.2002.1003567

[26] P. Ranganathan, V. S. Pai, and S. V. Adve. Using Speculative Retirement and Larger Instruction Windows to Narrow the Performance Gap between Memory Consistency Models. In

Proceedings of the Ninth ACM Symposium on Parallel Algorithms and Architectures, pp. 199–210, June 1997.

[27] O. Serlin. Fault-Tolerant Systems in Commercial Applications. *IEEE Computer*, pp. 19–30, Aug. 1984.

[28] T. J. Slegel et al. IBM's S/390 G5 Microprocessor Design. *IEEE Micro*, pp. 12–23, March/ April 1999. doi:10.1109/40.755464

[29] J. E. Smith and A. R. Pleszkun. Implementing Precise Interrupts in Pipelined Processors. *IEEE Transactions on Computers*, C-37(5), pp. 562–573, May 1988. doi:10.1109/12.4607

[30] D. J. Sorin, M. M. Martin, M. D. Hill, and D. A. Wood. SafetyNet: Improving the Availability of Shared Memory Multiprocessors with Global Checkpoint/Recovery. In *Proceedings of the 29th Annual International Symposium on Computer Architecture*, pp. 123–134, May 2002. doi:10.1109/ISCA.2002.1003568

[31] F. Sultan, T. Nguyen, and L. Iftode. Scalable Fault-Tolerant Distributed Shared Memory. In *Proceedings of the 2000 ACM/IEEE Conference on Supercomputing*, Nov. 2000.

[32] Y. M. Wang, E. Chung, Y. Huang, and E. Elnozahy. Integrating Checkpointing with Transaction Processing. In *Proceedings of the 27th International Symposium on Fault-Tolerant Computing Systems*, pp. 304–308, June 1997. doi:10.1109/FTCS.1997.614103

[33] Y.-M. Wang, Y. Huang, K.-P. Vo, P.-Y. Chung, and C. Kintala. Checkpointing and Its Applications. In *Proceedings of the 25th International Symposium on Fault-Tolerant Computing Systems*, pp. 22–31, June 1995.

[34] K. Wu, W. K. Fuchs, and J. H. Patel. Error Recovery in Shared Memory Multiprocessors Using Private Caches. *IEEE Transactions on Parallel and Distributed Systems*, 1(2), pp. 231–240, Apr. 1990. doi:10.1109/71.80134

[35] K.-L. Wu and W. K. Fuchs. Recoverable Distributed Shared Virtual Memory. *IEEE Transactions on Computers*, 39(4), pp. 460–469, Apr. 1990. doi:10.1109/12.54839

•   •   •   •

CHAPTER 4

Diagnosis

In the past two chapters, we have discussed how to detect errors and recover from them. For transient errors, detection and recovery are sufficient. After recovery, the transient error is no longer present and execution can resume without a problem. However, if an error is due to a permanent fault, detection and recovery may not be sufficient.

In this chapter, we first present general concepts in diagnosis (Section 4.1), before delving into diagnosis schemes that are specific to microprocessor cores (Section 4.2), caches and memory (Section 4.3), and multiprocessors (Section 4.4). We conclude (Section 4.5) with a discussion of open challenges in this area.

4.1 GENERAL CONCEPTS

In this section, we motivate the use of diagnosis hardware (Section 4.1.1), explain why the difficulty of providing diagnosis depends heavily on the system model (Section 4.1.2), and present built-in self-test (BIST), which is one of the most common ways of performing diagnosis (Section 4.1.3).

4.1.1 The Benefits of Diagnosis

For processors with backward error recovery, just using error detection and recovery for fault tolerance could lead to livelock. If, after backward error recovery, the processor's execution keeps reencountering errors due to a permanent fault, then it will keep recovering and fail to make forward progress. For example, consider a permanent fault in a core's lone multiplier. If the fault is exercised by a given pair of input operands, then the core will detect the error and recover. After recovery to a pre-error recovery point (i.e., before the erroneous multiplication instruction), it will resume execution and eventually reach the same multiplication instruction again. Because the fault is permanent, the result will be erroneous again. The error will be detected and the core will recover again.

For processors that use forward error recovery, there is also a possible problem with simply using detection and recovery to tolerate permanent faults. The detection and FER schemes are designed for a specific error model, say one stuck-at error at a time. In the presence of a single permanent fault, the detection and FER schemes operate as expected. However, they can no longer tolerate any additional errors. The single-error model is not viable or realistic if latent errors due to permanent faults are not cleared from the system.

Thus, for processors with backward error recovery (BER) or forward error recovery (FER), it would be beneficial to be able to *diagnose* a permanent fault—determine that a permanent fault exists and isolate the location of the fault—so that the processor could repair itself. Self-repair, which is the subject of Chapter 5, involves using redundant hardware to replace hardware that has been diagnosed as permanently faulty. After diagnosis and self-repair, a processor with BER could make forward progress and a processor with FER would be rid of latent errors that invalidate its error model.

4.1.2   System Model Implications

The ease of gathering diagnostic information depends heavily on the system model. We divide this discussion into two parts: system models to which it is easy to add diagnosis and system models to which adding diagnosis is challenging.

"Easy" System Models. Processors that use forward error recovery get fault isolation (i.e., the location of the fault) for free. To correct an error, without recovering to a prior state, requires the processor to know where the error is so that it can be fixed. For example, if a triple modular redundancy (TMR) system has a fault in one of its three modules, the voter will identify which of the modules was outvoted by the other two. The outvoted module had the error. Similarly, for an error-correcting code (ECC) to produce the correct data word from an erroneous codeword, it must know which bits in the erroneous codeword contained the errors.

Like processors that use FER, processors that use BER in conjunction with localized (i.e., not end-to-end) error detection schemes also get fault isolation for free. For example, if errors in a multiplier are detected by a dedicated modulo checker, then the modulo checker provides diagnosis capability at the granularity of the multiplier. The granularity of the diagnosis is equal to the granularity at which errors are detected.

Thus, a system with FER or a system with localized error detection schemes knows where an error is, but it does not know if the error is transient or due to a permanent fault. A simple way to determine if a permanent fault exists in a module is to maintain an error counter for that module. If more than a predefined threshold of errors is observed within a predefined window of time, then the system assumes that the errors are due to a permanent fault. A permanent fault is likely to lead to many errors in a short amount of time. Using an error counter in this way enables the system to correctly ignore transient errors, because transient errors occur relatively infrequently.

"Hard" System Models. From an architect's point of view, the most challenging system model for diagnosis is a system with BER and end-to-end error detection. End-to-end error detection schemes, which detect errors at a high level, often provide little or no diagnostic information. For example, the SWAT [6] error detection scheme uses the occurrence of a program crash as one of its indicators of an error. If an error is detected in this fashion, it is impossible to know *why* the system crashed.

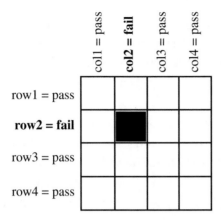

FIGURE 4.1: Using BIST for diagnosis. Assume that the BIST hardware tests each row and each column. Based on which tests pass and fail, the BIST hardware can identify the faulty component(s). In this case, the tests for row 2 and column 2 fail, indicating that the shaded entry is faulty.

In this system model, the architect must add dedicated diagnostic hardware to the system or suffer the problems discussed in Section 4.1.1. Because adding diagnosis to the other system models is so straightforward, we focus in the rest of this chapter on systems with BER and end-to-end error detection.

4.1.3 Built-In Self-Test
One common, general form of diagnostic hardware is BIST. BIST hardware generates test inputs for the system and compares the output of the system to a prestored, known-good output for that set of test inputs. If the system produces outputs that differ from the expected outputs, the system has at least one permanent fault. Often, the differences between a system's outputs and the expected outputs provide diagnostic information. Figure 4.1 illustrates an example in which BIST is used to diagnose faults in an array of memory cells. BIST hardware is often invoked when a system is powered on, but it can also be used at other times to detect permanent faults that occur in the field.

4.2 MICROPROCESSOR CORE
As the threat of permanent faults has increased, there has been a recent surge in research into diagnosis for microprocessor cores.

4.2.1 Using Periodic BIST
A straightforward diagnosis approach is to periodically use BIST. BulletProof [9] performs periodic BIST of every component in the core. During each "computation epoch," which is the time between

taken checkpoints, the core uses spare cycles to perform BIST (e.g., testing the adder when the adder would otherwise be idle). If BIST identifies a permanent fault, then the core recovers to a prior checkpoint. If BIST does not identify any permanent faults, then the computation epoch was executed on fault-free hardware and a checkpoint can be taken that incorporates the state produced during that epoch.

Constantinides et al. [3] showed how to increase the flexibility and reduce the hardware cost of the BulletProof approach by implementing the BIST partially in software. Their scheme adds instructions to the ISA that can access and modify the scan chain used for BIST; using these instructions, test programs can be written that have the same capability as all-hardware BIST.

Similar to BulletProof, FIRST [10] uses periodic BIST, but with two important differences. First, the testing is intended to detect emerging wear-out faults. As wear-out progresses, a circuit is likely to perform more slowly and thus closer to its frequency guardband. FIRST tests circuits closer to their guardbands to detect wear-out before the circuit fails completely (i.e., exceeds its frequency guardband). Second, because wear-out occurs over long time spans, the interval between tests is far longer, on the order of once per day.

4.2.2 Diagnosing During Normal Execution

Instead of adding hardware to generate tests and compare them to known outputs, another option is to diagnose faults as the core is executing normally. An advantage of this scheme, compared to BIST, is that it can achieve lower hardware costs.

Bower et al. [1] use a statistical diagnosis scheme for diagnosing permanent faults in superscalar cores. They assume that the core has an end-to-end error detection mechanism that detects errors at an instruction granularity (e.g., redundant multithreading or DIVA). This form of error detection, by itself, provides little help in diagnosis. They add an error counter for each unit that is potentially diagnosable, including ALUs, registers, reorder buffer entries, and so on. During execution, each instruction remembers which units it used. If the error detection mechanism detects an error in an instruction, it uses BER to recover from the error and it increments the error counters for each unit used by that instruction. If instructions are assigned to units in a fairly uniform fashion, then the error counter of a unit with a permanent fault will get incremented far more quickly than the error counter for any other unit. If an error counter exceeds a predefined threshold within a predefined window of time, then the unit associated with that error counter is diagnosed as having a permanent fault. If the core has only a singleton instance of unit X and a singleton instance of unit Y, and both unit X and unit Y are used by all instructions, then a permanent fault in either unit is indistinguishable from a permanent fault in the other. This limitation of the diagnosis scheme may not matter, though, because a permanent fault in any singleton unit is unrepairable; knowing which singleton unit is faulty is not helpful.

Li et al. [5] developed a diagnosis scheme that works in conjunction with an even higher level end-to-end detection mechanism. Their work assumes that errors are detected when they cause anomalous software behavior, using SWAT [6] (discussed in Section 2.2.6). This form of error detection provides virtually no diagnosis information. If an anomalous behavior, such as a program crash, is detected, the core uses BER to recover to a pre-error recovery point and enters a diagnosis mode. During diagnosis, the pre-error checkpoint is copied to another core that is assumed to be fault-free. These two cores then both execute from the pre-error checkpoint and generate execution traces that are saved. By comparing the two traces and analyzing where they diverge, the diagnosis scheme can diagnose the permanent fault with excellent accuracy.

4.3 CACHES AND MEMORY

As we will explain in Chapter 5, for caches and memories, the most common granularity of self-repair is the row or column. Storage structures are arranged as arrays of bits, and self-repair is more efficient for rows and columns than for individual bits or arbitrary groups of bits. Thus, the goal of diagnosis is to identify permanently faulty rows and columns.

The primary approach for cache and memory diagnosis is BIST, and this well-studied approach has been used for decades [8, 12]. The BIST unit generates sequences of reads and writes to the storage structure and, based on the results, can identify permanently faulty rows and columns. These test sequences are often sophisticated enough to diagnose more than just stuck-at faults; in particular, many BIST algorithms can diagnose faults that cause the values on neighboring cells to affect each other.

Another potential approach to cache and memory diagnosis is ECC. As mentioned in Section 4.1.2, ECC must implicitly diagnose the erroneous bits to correct them. However, because the granularity of ECC is almost always far finer than that of the self-repair, ECC is not commonly used for explicit diagnosis (i.e., to guide self-repair).

4.4 MULTIPROCESSORS

Many traditional, multichip multiprocessors have had dedicated hardware for performing diagnosis. Particularly for systems with hundreds and even thousands of processors, there is a significant probability that some components (cores, switches, links, etc.) are faulty. Without hardware support for diagnosis, the system administrators would have a miserable time performing diagnosis and system availability would be low. We now discuss three well-known multiprocessors that provide hardware support for diagnosis.

The Connection Machine CM-5 [4] provides an excellent example of a supercomputer that provides substantial diagnostic capability. The CM-5 dedicates a processor to controlling the diagnostic tests, and it dedicates an entire network for use during diagnosis. The diagnosis network provides "back door" access to components, and the diagnostic tests use this access to isolate which components are faulty.

IBM's zSeries mainframes [7, 11] provide extensive diagnosis capabilities. By detecting errors soon after faults occur, mainframes prevent errors from propagating far from their origin and thus minimize how many components could contribute to any detected error. Mainframes also keep detailed error logs and process these logs to infer the existence and location of permanent faults.

Sun Microsystems's UltraEnterprise E10000 [2] dedicates one processor as the system service processor (SSP). The SSP is responsible for performing diagnostic tests and reconfiguring the system in response to faulty components.

4.5   CONCLUSIONS

Fault diagnosis is a reemerging area of research. After a long history of heavy-weight fault diagnosis in mainframes and supercomputers, low-cost fault diagnosis just recently emerged as a hot research topic in the computer architecture community. There are still numerous open problems to be solved, including the following two:

- Diagnosing faults in the memory system: We know how to diagnose faults in cores, caches, and memories, but diagnosing faults in the other components of a processor's memory system remains a challenge. These components include cache controllers, memory controllers, and the interconnection network. A related question, particularly for controllers, is how to develop self-repair schemes for these components. As we discuss in Chapter 5, self-repair for these components is also an open problem.

- Diagnosis granularity: It is not yet entirely clear what is an appropriate granularity for diagnosis (and self-repair). Furthermore, the choice of granularity depends on the expected number of permanent faults and the desired lifetime of the processor. The same granularity is unlikely to be appropriate for both a high-performance laptop processor and a processor that is embedded in a car.

4.6   REFERENCES

[1]   F. A. Bower, D. J. Sorin, and S. Ozev. A Mechanism for Online Diagnosis of Hard Faults in Microprocessors. In *Proceedings of the 38th Annual IEEE/ACM International Symposium on Microarchitecture*, pp. 197–208, Nov. 2005. doi:10.1109/MICRO.2005.8

[2]   A. Charlesworth. Starfire: Extending the SMP Envelope. *IEEE Micro*, 18(1), pp. 39–49, Jan./Feb. 1998.

[3]   K. Constantinides, O. Mutlu, T. Austin, and V. Bertacco. Software-Based Online Detection of Hardware Defects: Mechanisms, Architectural Support, and Evaluation. In *Proceedings of the 40th Annual IEEE/ACM International Symposium on Microarchitecture*, pp. 97–108, Dec. 2007.

[4]   C. E. Leiserson et al. The Network Architecture of the Connection Machine CM-5. In *Proceedings of the Fourth ACM Symposium on Parallel Algorithms and Architectures*, pp. 272–285, June 1992. doi:10.1145/140901.141883

[5]   M.-L. Li, P. Ramachandran, S. K. Sahoo, S. Adve, V. Adve, and Y. Zhou. Trace-Based Diagnosis of Permanent Hardware Faults. In *Proceedings of the International Conference on Dependable Systems and Networks*, June 2008.

[6]   M.-L. Li, P. Ramachandran, S. K. Sahoo, S. Adve, V. Adve, and Y. Zhou. Understanding the Propagation of Hard Errors to Software and Implications for Resilient System Design. In *Proceedings of the Thirteenth International Conference on Architectural Support for Programming Languages and Operating Systems*, Mar. 2008. doi:10.1145/1346281.1346315

[7]   M. Mueller, L. Alves, W. Fischer, M. Fair, and I. Modi. RAS Strategy for IBM S/390 G5 and G6. *IBM Journal of Research and Development*, 43(5/6), Sept./Nov. 1999.

[8]   R. Rajsuman. Deisgn and Test of Large Embedded Memories: An Overview. *IEEE Design & Test of Computers*, pp. 16–27, May/June 2001.

[9]   S. Shyam, K. Constantinides, S. Phadke, V. Bertacco, and T. Austin. Ultra Low-Cost Defect Protection for Microprocessor Pipelines. In *Proceedings of the Twelfth International Conference on Architectural Support for Programming Languages and Operating Systems*, Oct. 2006. doi:10.1145/1168857.1168868

[10] J. C. Smolens, B. T. Gold, J. C. Hoe, B. Falsafi, and K. Mai. Detecting Emerging Wearout Faults. In *Proceedings of the Workshop on Silicon Errors in Logic—System Effects*, Apr. 2007.

[11] L. Spainhower and T. A. Gregg. IBM S/390 Parallel Enterprise Server G5 Fault Tolerance: A Historical Perspective. *IBM Journal of Research and Development*, 43(5/6), Sept./Nov. 1999.

[12] R. Treuer and V. K. Agarwal. Built-In Self-Diagnosis for Repairable Embedded RAMs. *IEEE Design & Test of Computers*, pp. 24–33, June 1993. doi:10.1109/54.211525

CHAPTER 5

Self-Repair

In Chapter 4, we discussed how to diagnose permanent faults. Diagnosis, by itself, is not useful, though. Diagnosis is useful when it is combined with the ability of a processor to repair itself. In this chapter, we discuss some of the many ways in which a processor can perform self-repair. The unifying theme to all self-repair schemes is that they require physical redundancy. Without physical redundancy, no self-repair is possible.

5.1   GENERAL CONCEPTS

Fundamentally, self-repair involves physical redundancy and reconfiguration. If component A is diagnosed as permanently faulty, then the processor reconfigures itself to use component B instead of component A. Component A and component B are often homogeneous, but heterogeneity is also possible. For example, consider a processor with a complex ALU and a simple ALU. If the simple ALU fails, then the complex ALU can be used to perform the operations that the simple ALU would have otherwise performed.

Component B might be a "cold spare" that was not being used, or it might be a "hot spare" that was being used in conjunction with component A. Cold spares use no power and suffer little or no wear-out until they are enabled. However, a cold spare is effectively useless hardware until it is enabled. A cold spare may also need to be warmed up before it can begin operation. For example, consider a system with a cold spare core. For the cold spare core to take over for a faulty core, the system would need to first transfer a prefault recovery point from the faulty core to the cold spare.

Another important design issue for self-repair is the granularity at which the processor can repair itself. The only fundamental restriction is that the granularity of self-repair must be at least as coarse as the granularity of diagnosis. If the diagnosis scheme can only resolve that the ALU is faulty, then being able to repair just the adder within the ALU is not useful. The granularity of self-repair has several important ramifications. A coarser-grain self-repair requires less reconfiguration hardware and is simpler to implement. However, a coarser-grain self-repair may waste a significant amount of fault-free hardware. For example, if self-repair is performed at a core granularity, then a core with one single permanently faulty transistor is unusable although the millions of other transistors are fault-free. One might then conclude that finer-grain self-repair is preferable, but this conclusion is true only up to a certain point. At an extreme, one could imagine self-repair at the granularity of an individual logic gate. This absurdly fine granularity of self-repair would require more hardware for reconfiguration than the amount of hardware in the original design.

5.2   MICROPROCESSOR CORES

The approach to providing self-repair within a core depends heavily on the type of core. A superscalar core already has quite a bit of redundancy that can be exploited, and adding a bit more redundancy to a superscalar core may also be feasible. However, the simple, power-efficient cores that are being used in processors like Sun's UltraSPARC T1 [9] and T2 [17] and Intel's Larrabee [16] have little inherent redundancy. Adding a significant amount of redundant hardware to one of these cores is likely to defeat the goal of keeping the core small and simple.

5.2.1   Superscalar Cores

Superscalar cores naturally have redundancy. To execute multiple instructions per cycle requires replicas of many components, including ALUs and other functional units. Superscalar cores also contain array structures that have more entries than strictly necessary, such as the number of physical registers, reorder buffer entries, and load-store queue entries.

Shivakumar et al. [18] observed that the existing redundancy within a superscalar core is often sufficient to overcome many possible manufacturing defects (i.e., permanent faults introduced during chip fabrication). They introduce the notion of "performance averaged yield," which is similar to the usual chip yield metric but is scaled by the performance of the chips. A fully functional chip still has a yield of one. A chip that is still functional but has X% of the performance of a fully functional chip due to a few faulty components would have a yield of X%. The performance averaged yield metric is the aggregate yield over the total number of fabricated chips. By using the existing intracore redundancy, chips with faults can still contribute to the performance-averaged yield.

Srinivasan et al. [20], like Shivakumar et al. [18], seek to exploit the existing intracore redundancy for fault tolerance, and they make two important contributions beyond this prior work. They explore the possibility of adding cold spare components, and they consider permanent faults that occur in the field, instead of just manufacturing defects. Their results show that being able to exploit inherent and added redundancy can dramatically increase the lifetime reliability of a core.

Bower et al. [3] developed self-repairing array structures (SRAS) that can tolerate faults in the many buffers and tables in a typical superscalar core. These structures include the reorder buffer, reservation stations, and branch history table. For circular buffers, the repair logic involves a fault map to record which entries are faulty and some additions to the head and tail pointer advancement logic. For randomly addressable tables, the repair logic includes a fault map and remap logic that allows an access to a faulty entry to be steered to a spare, fault-free entry.

Schuchman and Vijaykumar's Rescue microarchitecture [13], like the work by Shivakumar et al. [18], seeks to improve a superscalar core's resilience to manufacturing defects. The goal of Rescue is to be a testable and defect-tolerant core. To achieve these goals, Rescue is designed to be easily divisible into "ways." A k-wide superscalar core consists of k ways that are not usually easy to isolate, but Rescue intentionally isolates the ways so that a defect in one way does not affect any other ways. Rescue's granularity of self-repair is a way, which is far coarser than the granularity in the other schemes discussed in this section.

5.2.2 Simple Cores

Unlike superscalar cores, simple cores have little inherent redundancy. A simple core is likely to have only one ALU, one multiplier, and so on. If a component is faulty, there are only two possible means of self-repair. The first approach is to use software support to reconstruct the functionality of the faulty component, and we discuss two implementations of this approach in this section. The other approach is to borrow this functionality from other cores on the chip, and we discuss this option when we present multiprocessor self-repair in Section 5.4.

Joseph's [8] Core Salvage scheme uses a virtual machine monitor (VMM) that detects when an instruction needs to use a faulty functional unit. The VMM then emulates the faulty functional unit instead of allowing the instruction to use the faulty functional unit. Many functional units are simple to emulate. In fact, there is a long history of emulating, rather than implementing, floating point units in low-cost cores. This previous emulation was done to reduce costs, but now it is being used for self-repair.

Meixner and Sorin's [10] Detouring scheme modifies the compiler to take a fault map as an additional input and produce a binary executable that touches none of the faulty hardware. The compiler can "Detour" around permanent faults in functional units, like Core Salvage, and it can also Detour around permanent faults in registers, instruction cache frames, and operand bypass network paths. The Detouring compiler can Detour around a faulty register by simply not allocating it. The compiler can Detour around instruction cache frames through careful layout of the code. Bypass network faults can be Detoured by either switching which operand uses which bypass path or by inserting NOPs to avoid using a path completely.

5.3 CACHES AND MEMORY

Storage structures are large and consist of many identical components. Because of these two features, being able to repair them is both important and fairly straightforward. The key to self-repair is to provide some number of spare storage cells and use them to replace faulty cells. The engineering challenges are determining how many spares to provide and at what granularity to perform the self-repair.

For SRAM caches and DRAM memory chips, there is a long history of using spare rows and columns [5, 2, 14, 11, 2]. Because of the array-like layout of storage structures, performing self-repair at the row and column granularity is far easier than repairing arbitrary groups of bits. Rows and columns are also at a "sweet spot" in terms of size. Self-repair of larger groups of bits is apt to waste many fault-free bits within the repaired group, and self-repair of smaller groups of bits requires significantly more circuitry for reconfiguration.

The only other "sweet spot" for self-repair is at the DRAM chip granularity. There is a wide range of faults that can cause most or all of a DRAM chip to be unusable. To address this fault model, many highly available systems, including IBM's S/390 Parallel Enterprise Server G5 [19], provide spare DRAM chips.

Recently, there has been renewed interest in cache self-repair because of the desire to drastically reduce the cache's supply voltage. Reducing the voltage reduces power consumption, but it also makes many of the cells unusable. Architects are trying to find appropriate trade-offs between reducing the power consumption and increasing the number of faulty cells. Wilkerson et al. [21] developed two schemes for tolerating large numbers of faulty cache cells and thus enabling voltage reductions. One scheme uses a quarter of the cache to store the fault locations and repair bits for the rest of the cache. The other scheme uses pairs of cache lines to form logical cache lines; the fault-free bits of each pair of lines are used to implement a single logical line.

5.4   MULTIPROCESSORS

Having multiple cores provides more opportunity for self-repair because there is inherently much more redundancy on the chip. One question for architects is how best to use this inherent redundancy. Another question is whether to add even more redundancy for certain noncore components of the chip.

5.4.1   Core Replacement

A straightforward approach to multicore self-repair is to simply disable a faulty core and replace its functionality with either a cold spare core or one of the other currently running cores. This approach adds little complexity, and it has been adopted by researchers [1], IBM mainframes [19], and commercial processors like the Cell Broadband Engine [6]. Sony's PlayStation 3 explicitly uses only seven of the eight synergistic processing element (SPE) cores in the Cell processor to be able to accommodate the significant fraction of Cell processors that have faults in one SPE. For processors with many cores and few expected permanent faults, performing self-repair at the core granularity is a reasonable solution.

5.4.2   Using the Scheduler to Hide Faulty Functional Units

Joseph's [8] Core Salvage scheme, which we first discussed in Section 5.2, also presents an appealing solution for using the multiple cores in a processor for purposes of self-repair. The key idea is to match threads to the cores that have the fault-free hardware necessary to run them. Consider the example in Figure 5.1. Assume that thread T1 heavily uses the multiplier but never uses the divider. Assume that thread T2 heavily uses the divider but rarely uses the multiplier. Assume also that T1 is initially running on core C1, which has a faulty multiplier, and T2 is running on core C2, which has a faulty divider. In this situation, migrating the threads such that T1 runs on C2 and T2 runs on C1 is beneficial. C1 will still need to emulate a few multiplication instructions, using the emulation technique described in Section 5.2, but the overall performance of the multicore processor is nearly that of a fault-free multicore.

5.4.3   Sharing Resources Across Cores

A more general approach to multicore self-repair is to cobble together the necessary fault-free resources from among multiple cores. Particularly, if the cores are simple and have little or no intracore redundancy, then sharing resources across cores is an attractive solution. The engineering challenges are determining the granularity at which resources should be shared and then how exactly to share them. In this section, we discuss two similar approaches for self-repair of multicore processors that use only simple cores. The developers of both approaches determined that enabling self-repair at the granularity of a pipeline stage is a "sweet spot" in terms of fault tolerance, performance, and design complexity.

Gupta et al.'s [7] StageNet architecture provides full reconfigurability among pipeline stages. Consider a baseline processor with some number of cores. As illustrated in Figure 5.2, StageNet adds a crossbar between every set of pipeline stages. These crossbars enable the processor to be organized into a set of logical cores that consist of stages from multiple physical cores. For example, if core 1 has a faulty execute stage, it can borrow the execute stage of one of the other cores to create a fully functional logical core.

Romanescu and Sorin's [12] Core Cannibalization Architecture (CCA) is similar to StageNet, except that it replaces the crossbars with dedicated wires between neighboring stages. CCA

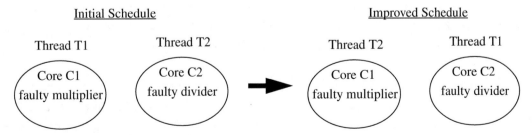

FIGURE 5.1: Example of using scheduler to hide faulty functional units. Assume that thread T1 heavily uses the multiplier and T2 heavily uses the divider. Switching the mapping of threads to cores greatly improves processor performance.

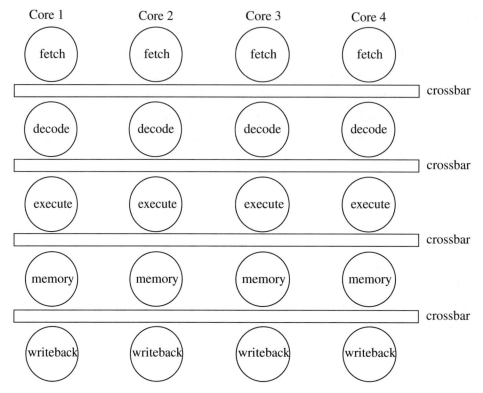

FIGURE 5.2: High-level illustration of StageNet [7].

provides less reconfigurability, but it achieves lower performance overheads by avoiding the latencies through the crossbars.

5.4.4 Self-Repair of Noncore Components

One well-studied aspect of multiprocessor self-repair is self-repair of the interconnection network. Certain network topologies lend themselves to self-repair because they contain multiple paths between cores. If there is a permanent fault along one path, then communication can use one of the other paths.

Many commercial machines have provided the ability to map out faulty portions of the interconnection network, and we present two well-known examples here. The Cray T3D and T3E [15] use a 3D torus topology, which provides an ample number of minimum length paths between pairs of cores. The Sun UltraEnterprise E10000 [4] has four broadcast buses that are interleaved by address. If one bus fails, then the system maps out this bus, and it also maps out another bus to keep the address interleaving simple.

Aggarwal et al. [1] recently proposed a multicore chip design that provides self-repair capability to avoid scenarios in which a single permanent fault either renders the chip unusable or drastically reduces its performance. Their design efficiently provides redundancy for memory controllers, link adapters, and interconnection network resources, such that the redundant resources can be shared among the cores whenever possible.

5.5   CONCLUSIONS

The issue of self-repair is closely tied to the issue of diagnosis, and there are thus many similarities in their current statuses and open problems. We foresee at least two promising areas of research in this area:

- Expanding software self-repair: Using software for self-repair, as in Core Salvage [8] or Detouring [10], is appealing, but it currently offers a limited amount of coverage. There are many faults that cannot be repaired using current software self-repair schemes. A breakthrough in this area, perhaps using some hardware support, would be a significant contribution.
- Tolerating a greater number of permanent faults: Many of the schemes we have presented in this section are appropriate for only small numbers of permanent faults. StageNet [7] and CCA [12], discussed in Section 5.4.3, can tolerate a handful of permanent faults, but they cannot tolerate dozens of permanent faults. To tolerate more faults requires a finer granularity of self-repair, and it is an open problem how to create an efficient self-repair scheme at a much finer granularity. As discussed in Chapter 4, the granularity of self-repair is tied to the granularity of diagnosis, so a finer-grained self-repair will require a finer-grained diagnosis. One possibly fundamental roadblock in this area of research is that the additional hardware added to enable self-repair is also vulnerable to permanent faults. At some point, adding more hardware may actually *reduce* the dependability of the processor.

5.6   REFERENCES

[1]   N. Aggarwal, P. Ranganathan, N. P. Jouppi, and J. E. Smith. Configurable Isolation: Building High Availability Systems with Commodity Multi-Core Processors. In *Proceedings of the 34th Annual International Symposium on Computer Architecture*, pp. 470–481, June 2007.

[2]   D. K. Bhavsar. An Algorithm for Row-Column Self-Repair of RAMs and Its Implementation in the Alpha 21264. In *Proceedings of the International Test Conference*, pp. 311–318, 1999. doi:10.1109/TEST.1999.805645

[3]   F. A. Bower, S. Ozev, and D. J. Sorin. Autonomic Microprocessor Execution via Self-Repairing Arrays. *IEEE Transactions on Dependable and Secure Computing*, 2(4), pp. 297–310, Oct.-Dec. 2005. doi:10.1109/TDSC.2005.44

[4]   A. Charlesworth. Starfire: Extending the SMP Envelope. *IEEE Micro*, 18(1), pp. 39–49, Jan./Feb. 1998.

[5]   T. Chen and G. Sunada. A Self-Testing and Self-Repairing Structure for Ultra-Large Capacity Memories. In *Proceedings of the International Test Conference*, pp. 623–631, Oct. 1992. doi:10.1109/TEST.1992.527883

[6]   M. Gschwind et al. Synergistic Processing in Cell's Multicore Architecture. *IEEE Micro*, 26(2), pp. 10–24, Mar./Apr. 2006.

[7]   S. Gupta, S. Feng, A. Ansari, J. Blome, and S. Mahlke. The StageNet Fabric for Constructing Reslilient Multicore Systems. In *Proceedings of the 41st Annual IEEE/ACM International Symposium on Microarchitecture*, pp. 141–151, Nov. 2008.

[8]   R. Joseph. Exploring Core Salvage Techniques for Multi-core Architectures. In *Proceedings of the Workshop on High Performance Computing Reliability Issues*, Feb. 2005.

[9]   P. Kongetira, K. Aingaran, and K. Olukotun. Niagara: A 32-Way Multithreaded SPARC Processor. *IEEE Micro*, 25(2), pp. 21–29, Mar./Apr. 2005. doi:10.1109/MM.2005.35

[10]   A. Meixner and D. J. Sorin. Detouring: Translating Software to Circumvent Hard Faults in Simple Cores. In *Proceedings of the International Conference on Dependable Systems and Networks*, June 2008.

[11]   R. Rajsuman. Design and Test of Large Embedded Memories: An Overview. *IEEE Design & Test of Computers*, pp. 16–27, May/June 2001. doi:10.1109/54.922800

[12]   B. F. Romanescu and D. J. Sorin. Core Cannibalization Architecture: Improving Lifetime Chip Performance for Multicore Processors in the Presence of Hard Faults. In *Seventeenth International Conference on Parallel Architectures and Compilation Techniques*, Oct. 2008.

[13]   E. Schuchman and T. N. Vijaykumar. Rescue: A Microarchitecture for Testability and Defect Tolerance. In *Proceedings of the 32nd Annual International Symposium on Computer Architecture*, pp. 160–171, June 2005. doi:10.1109/ISCA.2005.44

[14]   S. E. Schuster. Multiple Word/Bit Line Redundancy for Semiconductor Memories. *IEEE Journal of Solid-State Circuits*, SC-13(5), pp. 698–703, Oct. 1978. doi:10.1109/JSSC.1978.1051122

[15]   S. L. Scott. Synchronization and Communication in the Cray T3E Multiprocessor. In *Proceedings of the Seventh International Conference on Architectural Support for Programming Languages and Operating Systems*, pp. 26–36, Oct. 1996.

[16]   L. Seiler et al. Larrabee: A Many-Core x86 Architecture for Visual Computing. In *Proceedings of ACM SIGGRAPH*, 2008.

[17]   M. Shah et al. UltraSPARC T2: A Highly-Threaded, Power-Efficient, SPARC SOC. In *Proceedings of the IEEE Asian Solid-State Circuits Conference*, pp. 22–25, Nov. 2007.

[18] P. Shivakumar, S. W. Keckler, C. R. Moore, and D. Burger. Exploiting Microarchitectural Redundancy For Defect Tolerance. In *Proceedings of the 21st International Conference on Computer Design*, Oct. 2003.

[19] L. Spainhower and T. A. Gregg. IBM S/390 Parallel Enterprise Server G5 Fault Tolerance: A Historical Perspective. *IBM Journal of Research and Development*, 43(5/6), Sept./Nov. 1999.

[20] J. Srinivasan, S. V. Adve, P. Bose, and J. A. Rivers. Exploiting Structural Duplication for Lifetime Reliability Enhancement. In *Proceedings of the 32nd Annual International Symposium on Computer Architecture*, June 2005. doi:10.1109/ISCA.2005.28

[21] C. Wilkerson et al. Trading off Cache Capacity for Reliability to Enable Low Voltage Operation. In *Proceedings of the 35th Annual International Symposium on Computer Architecture*, pp. 203–214, June 2008.

· · · ·

CHAPTER 6

The Future

This book represents a snapshot of the field as of January 2009. Fault-tolerant computer architecture is a vibrant field that has been reinvigorated in the past 10 years or so by forecasts of increasing fault rates, and we expect this field to evolve quite a bit in the upcoming years as the current reliability challenges become more acute and new challenges arise. The general concepts described in this book will not become obsolete, but we expect (and hope!) that many new ideas and implementations will be developed to address current and emerging challenges. In the four main chapters of this book, we have identified some of the open problems to be solved, and we anticipate that those problems, as well as problems that have not even arisen yet, will be tackled.

6.1 ADOPTION BY INDUSTRY

Despite the recent excitement about research in fault-tolerant computer architecture, few of the products of this renaissance of research have thus far found their way into commodity processors. Industry is understandably reluctant to add anything seemingly complicated or costly until absolutely required, and current fault rates have not yet led to enough user-visible hardware failures to persuade much of the industry that sophisticated fault tolerance is necessary. Industry has been willing to adopt fault tolerance mechanisms that provide a large "bang for the buck," such as adding low-cost parity to detect all single-bit errors in a cache, but more sophisticated and costly fault tolerance mechanisms have been confined to mainframes, supercomputers, and mission-critical embedded processors.

Nevertheless, despite industry's current reluctance to adopt fault tolerance techniques, industry is unlikely to be able to maintain that attitude. Fault rates are expected to increase dramatically in future generations of CMOS, and future nanotechnologies that may replace CMOS are expected to be even less reliable. Processors implemented in such technologies are unlikely to be dependable enough without substantial built-in fault tolerance. We are approaching the end of the era in which we could design a processor largely without thinking about faults and then, perhaps, we could add on parity bits or ECC after the design is mostly complete.

6.2   FUTURE RELATIONSHIPS BETWEEN FAULT TOLERANCE AND OTHER FIELDS

We are intrigued by what the future holds for the relationships between fault tolerance and many other aspects of system design. A few of the more interesting factors that are inter-dependent with fault tolerance are:

6.2.1   Power and Temperature

We have discussed how increasing power consumption leads to increasing temperatures, which then leads to decreases in reliability. For many years, new generations of microprocessors consumed ever-increasing amounts of power, but recently, architects have hit a so-called power wall. If anything, the amount of power consumed per processor may decrease due to the cost of power. There has also been a recent surge of research into thermal management [4], and there is a clear synergy between managing temperature and managing reliability.

6.2.2   Security

At a high level, a security breach is just another type of fault to be tolerated. However, the mechanisms used to tolerate these types of "faults" are often far different from those used to tolerate physical faults. Being able to integrate these two areas would be exciting, and some initial work has explored this possibility [3].

6.2.3   Static Design Verification

We have discussed mechanisms for tolerating errors due to design bugs, but researchers have not yet fully explored the relationship between static verification and runtime fault tolerance. We are intrigued by recent work that explicitly trades off which core design bugs are eliminated by static verification and which are detected by runtime hardware [5], and we look forward to future work in this area.

6.2.4   Fault Vulnerability Reduction

The development of the architectural vulnerability metric by Mukherjee et al. [2] has inspired a vast amount of work in analyzing and reducing hardware's vulnerability to faults. Analogous to our discussion of static design verification, we are curious to see how future research continues to integrate vulnerability reductions with runtime fault tolerance.

6.2.5   Tolerating Software Bugs

In this book, we have focused on tolerating hardware faults. One could argue, though, that software faults (bugs) are an equal or bigger problem. A system that tolerates hardware faults will execute a program exactly as it is written—and it will faithfully execute buggy software. Developing hardware that can help tolerate software bugs, perhaps by detecting anomalous behaviors, would be an important contribution. Some initial work [1, 6, 7] has been done, and we expect this area of research to remain active because of its importance.

6.3   REFERENCES

[1]   S. Lu, J. Tucek, F. Qin, and Y. Zhou. AVIO: Detecting Atomicity Violations via Access Interleaving Invariants. In *Proceedings of the Twelfth International Conference on Architectural Support for Programming Languages and Operating Systems*, Oct. 2006.

[2]   S. S. Mukherjee, C. Weaver, J. Emer, S. K. Reinhardt, and T. Austin. A Systematic Methodology to Compute the Architectural Vulnerability Factors for a High-Performance Microprocessor. In *Proceedings of the 36th Annual IEEE/ACM International Symposium on Microarchitecture*, Dec. 2003. doi:10.1109/MICRO.2003.1253181

[3]   N. Nakka, Z. Kalbarczyk, R. Iyer, and J. Xu. An Architectural Framework for Providing Reliability and Security Support. In *Proceedings of the International Conference on Dependable Systems and Networks*, June 2004. doi:10.1109/DSN.2004.1311929

[4]   K. Skadron, M. R. Stan, W. Huang, S. Velusamy, K. Sankaranarayanan, and D. Tarjan. Temperature-aware Microarchitecture. In *Proceedings of the 30th Annual International Symposium on Computer Architecture*, pp. 2–13, June 2003. doi:10.1145/859619.859620, doi:10.1145/859618.859620

[5]   I. Wagner and V. Bertacco. Engineering Trust with Semantic Guardians. In *Proceedings of the Design, Automation and Test in Europe Conference*, Apr. 2007.

[6]   E. Witchell, J. Cates, and K. Asanovic. Mondrian Memory Protection. In *Proceedings of the Tenth International Conference on Architectural Support for Programming Languages and Operating Systems*, pp. 304–316, Oct. 2002. doi:10.1145/605397.605429

[7]   P. Zhou, W. Liu, L. Fei, S. Lu, F. Qin, Y. Zhou, S. Midkiff, and J. Torrellas. AccMon: Automatically Detecting Memory-Related Bugs via Program Counter-based Invariants. In *Proceedings of the 37th Annual IEEE/ACM International Symposium on Microarchitecture*, pp. 269–280, Dec. 2004.

Author Biography

Daniel J. Sorin is an assistant professor of electrical and computer engineering and of computer science at Duke University. His research interests are in computer architecture, including dependable architectures, verification-aware processor design, and memory system design. He received his Ph.D. and M.S. in electrical and computer engineering from the University of Wisconsin and his B.S.E. in electrical engineering from Duke University. He is the recipient of an NSF Career Award and a Warren Faculty Scholarship at Duke University.

Printed in the United States
by Baker & Taylor Publisher Services